THE DARK SIDE OF THE BBC. A DIST(

PREFACE.

THE DARK SIDE OF THE BBC. A DISTURBING TRUE STORY.

THIS BOOK IS A TRUE STORY OF ACTUAL FACT. THERE IS NO FICTION INVOLVED WHATSOEVER. ALL OF THE CHARACTERS MENTIONED IN THIS BOOK ARE REAL PEOPLE THAT ACTUALLY EXIST. NO NAMES HAVE BEEN CHANGED. THIS STORY IS A REAL-LIFE EXPOSE OF THE ILLEGAL ACTIVITIES OF A TELEVISION COMPANY THAT EXISTS TODAY AND IS CALLED THE BBC (BRITISH BROADCASTING CORPORATION). IT IS A STORY THAT PROVES THE EXISTENCE AND MIS-USE OF TELEPATHY OR TELEPATHIC BROADCASTING (BROADCASTING DIRECTLY INTO PEOPLES MINDS) AND HOW IT IS DIRECTED AGAINST ORDINARY MEMBERS OF THE GENERAL PUBLIC IN ORDER TO CAUSE THEM TO SUFFER PARANOIA AND A NERVOUS BREAK-DOWN AS A MEANS OF 'FUN' FOR THOSE EVIL AND PERVERTED BBC PEOPLE OPERATING WITHIN THE 'DARK SIDE OF TELEVISION'. THIS ELECTRICALLY MANUFACTURED PHENONEMON HAS BEEN EXISTING AND CONTROLLING THE CITY OF BIRMINGHAM FOR OVER TWENTY YEARS NOW SINCE 1990 AND HAS BEEN LEFT UP TO THE MAIN VICTIM AND AUTHOR OF THIS BOOK (TONY HICKEY) TO OFFICIALLY EXPOSE AND ERADICATE. BUT THE BBC ARE A VERY EVIL AND INFLUENCIAL CORPORATION AND YOU WILL READ HOW THEY CORRUPT AND INTIMIDATE ANYONE TO HELP THEM COVER THIS STORY UP. APART FROM ME THAT IS. I DO NOT FEAR THEM. IN FACT, THEY FEAR ME. MY GRIM DETERMINATION MY INTELLIGENCE AND MY EXPLOSIVE VIOLENCE. IT IS TIME THAT EVERYONE KNEW THE REAL STORY OF WHAT EVEN THE BBC CALL THEIR 'DIRTY LITTLE SECRET'..... THE EXISTENCE OF TELEPATHY.

THE DARK SIDE OF THE BBC. A DISTURBING TRUE STORY.

THIS IS A TRUE STORY OF COURAGE AND PERSERVERENCE BY THE AUTHOR IN A DARK WORLD OF CRIME AND VIOLENCE CONTROLLED BY THE BRITISH BROADCASTING CORPORATION.

THE DARK SIDE OF THE BBC. A DISTURBING TRUE STORY.

THE DARK SIDE OF THE BBC. A DISTURBING TRUE STORY.

LONDON.

THE DARK SIDE OF THE BBC. A DISTURBING TRUE STORY.

THE DARK SIDE OF THE BBC. A DISTURBING TRUE STORY.

PATRICK'S FLAT.

Things weren't really working out for me at this time. I had found myself doing twelve months at Stafford Prison, where all the Midlands local boys end up. But really I was pissed off with myself. I'd lost nearly all my remission...stupid things really, like the night I feigned a suicide bid with the help of my two cell-mates, pretending I had slit my wrists at two o'clock in the morning. Wound the screws up nice, but I wished later I hadn't used so much of my strawberry jam as blood. I lost seven days remission, seven days loss of earnings and seven days down the block. My release date just kept coming real close and then getting put back further away again. You can probably imagine how this made me feel, I wasn't my usual, always buzzing, self. In fact I was so bad-tempered at times that pretty much everyone was keeping their distance from me this particular week. All except one of the con's I didn't mind associating with, Noel Connors from Birmingham, a fairly handy so and so when he wanted to be and who I knew from the out. We had some giggles in there, playing up most of the time out of sheer boredom, but a lot of the time I used to go into his cell during lock-up to thrash him at chess, only I never did manage to thrash him at all. Let's face it; the guy is shit hot when it comes to playing chess so I just couldn't

THE DARK SIDE OF THE BBC. A DISTURBING TRUE STORY.

beat him for any serious number of games. That just wasn't happening. But anyway, it was then that I had a visit from Patrick, my older brother from London. Now I did think that was a bit out of character at the time: He'd never before visited me on the inside, so when he turned up, I was gobsmacked. Gobsmacked but over the moon when he invited me down to his place in London. I asked what he was doing with himself out in the real world and he replied that he was still down South, in London, but was doing photography; models, no less.

Of course, he had a sample of his wares: a girl named Beccy, dark-haired, she had the face, the body, everything. Dressed tastefully as well, that lingerie must have cost a pretty penny. She was stunning. She actually reminded me of my second great love, Claire Jennings who I met when I was eighteen; she was from Druids Heath in Birmingham. My very first great love was called Janet Clarke and she was from Harborne in Birmingham. They would both later play a part in the mayhem that was to come a bit later in my life. So I returned to my cell feeling a little happier, but still pissed off about my situation. I served the rest of my sentence, looking at Beccy's picture nearly every day and before I'd even got out, my mind was made up: I was going to London for a new start, away from the crime and criminals, get a decent job, and maybe get off with Beccy.

I got a train ticket from Stafford to London from the prison when I was released. I was feeling my normal self again, smiling and happy, oozing confidence as I headed off for London. And that's how I started my new life in the big city. A reformed ex-con who didn't really give a shit about anything. A bit of a tasty, fun loving ex-con. I arrived at Patrick's flat about 2 o'clock in the

THE DARK SIDE OF THE BBC. A DISTURBING TRUE STORY.

afternoon. He'd just moved into a top flat on Girdler's Road in West Kensington. The place was nice but it was rather stark because he'd not managed to buy any furniture as yet. Neither of us was in the mood for sitting around indoors, so we were sampling the flavours of the West End within twenty minutes. I thought it was a bit odd how we managed to bump into at least one of Patrick's people in every pub and club we visited, and I had the sensation of being put on display as some kind of invader. A lot of the conversation seemed to get turned onto events in my past, a past I was really trying to forget. The thing was, in order to cultivate some (much needed) respect around his manor, Patrick had told everyone that his 'psycho' brother, fresh out of prison, was going to be around, and he didn't take kindly to dickheads or anyone trying to play one of his kindred. Now, I've always fought Patrick's battles in the past – he's my brother, and I look after my blood – but on reflection, I was slightly miffed that he didn't even ask me whether he could use my own reputation to prop up his public image. But the upshot was, I already had quite a reputation in London, thanks to Patrick, even before I arrived on the scene.

Within a matter of weeks of my arrival in London I got my first job working in the Charlie Chester Casino which is in Soho, the red light area of the West End. I was only a mere kitchen porter but the pay was over £200 per week and that was pretty good back in 1986. The head chef who hired me was a gay English chap who had one of those Filipino brides. The only reason he hired me was because of my madcap behaviour at the interview and he liked me. It was night work and I started the following Monday just before Christmas. This head chef would run the kitchen and openly encouraged his staff to talk and banter with me even though I was a lowly kitchen porter

THE DARK SIDE OF THE BBC. A DISTURBING TRUE STORY.

slaving over a dish filled sink. I suppose he wanted a bit of fun and banter in his kitchen with a hunky guy like me. Soon enough, I got to meet all the croupiers on their lunch hours and it wasn't long before I was the topic of conversation. The males seemed a bit wary of me (well lets face it, I do look like a bit of a fighter) but the females absolutely loved me, though I was a bit too new to the scene for them to get involved with as yet. I don"t think there was one single female who worked at the casino who wasn't an absolute stunner. Believe me they were all gorgeous and I couldn't wait to make a date with any one of them. Trouble was, for some unknown reason, other than their own paranoia, some of the males that worked within the casino floor would not integrate with me. I suppose they were a bit jealous of the fact that the womenfolk there had the hots for me and I had the hots for them. This came to light when one of the girls there stood up in the middle of the canteen and declared out loud "he can be back-up man". Back-up man I ask you! I could have taken over if I wanted to, but I just wanted to earn a living and maybe date the girls there. They were keen enough obviously but when Christmas arrived and with it the dreaded works Christmas party, I went and blew it didn't I?

It all started off reasonably well but I got a little tipsy after a couple of hours of drinking and dancing and when it came to the taking of photos bit, the girls decided to stand right in front of me to have their picture taken. They were all wearing really short ra-ra skirts so I reckoned they were just trying to tease me with those beautiful long legs and pert bums. I had to take some kind of action in the wind-up so I leaned forward in the chair I was sitting in and proceeded to lift one of the girls skirt from the back right up to the waistline, exposing a nice pair of brilliant white knickers to everyone behind me. Everyone was falling

THE DARK SIDE OF THE BBC. A DISTURBING TRUE STORY.

about with laughter, me included, as I sat there holding the skirt up for a good 20 seconds. That went down a treat, even some of the guys started coming up to me and making conversation after that but as the night wore on and the drink flowed I started to get really drunk and then it happened.

A group of us, about a dozen in all, left the party just as it was ending and went out nightclubbing around the west end. We were the Charlie Chester Casino guys and gals so we could get in anywhere for free. It was a good buzz clubbing it for free in the west end of London, I felt like royalty but after drinking so much booze I ended up walking around in a daze. Eventually I just completely blacked out and can't remember anything that happened after that. I didn't find out what happened that night until I went back to work the following evening. I could tell by the frosty atmosphere around me that something was amiss, no-one was talking to me or even acknowledging my existence for starters so I sort of knew I must have upset someone the night before after I'd blacked out and so I decided to approach one of the croupiers to find out the s p. As it happened, I was told that one of the girls had sat down to talk to me in the club last night and that I was really drunk. Then for some unknown reason I flew into a drunken tantrum and behaved quite nastily towards her, not actually violent , but apparently I was so scary that I left her in a state of shock and panic. I was told that she was a bit of a babe too so I really did feel bad about it and knew I just had to apologise.

A few hours passed by, then she came into the canteen with a friend where I was preparing sandwiches. I waited for a quiet moment, and then I approached her. I told her how sorry I was about last night and how I regretted it and how stupid it made me feel. When I had finished she smiled

THE DARK SIDE OF THE BBC. A DISTURBING TRUE STORY.

and said "that's OK Tony, apology accepted", then she went back to her seat and I went back to preparing sandwiches. After a couple of minutes I returned to where she was sitting for some reason or other, but she was gone! I looked at the table where she had been and saw a single, solitary Rolo, still wrapped in its open golden foil. I had to laugh to myself as I realised she had only gone and left me her last Rolo.

The Christmas party took place about two weeks before Christmas and it was about one week before Christmas when the gay head chef sacked me. It was because of what happened after the party I know, but he told me it was because I had been leaving work too early. Bullshit!
Everyone at the casino was to get a big Christmas hamper from Fordrough and Masons (a big posh department store in Knightsbridge) and an even bigger bonus in their pay packet, me included. But when the head chef sacked me he started ranting and raving that I was not going to get a hamper and that I was not going to get the bonus. He was only a little skinny fella so I was quite surprised, yet starting to get quite angry. My pay cheque had already been made up with the bonus, (which was about £200), but now he was telling me that he was going to take it back to the pay office and have the bonus deducted. I was fuming and there was an intensity in my eyes as I glared back at him. I was even contemplating punching him in the mouth but before I could do so the big boss man of the whole casino arrived on the scene and took a look at my pay cheque. I thought he was going to start as well but he didn't. Instead he turned to the head chef and started ranting and raving at him! I thought go on you beauty, give it to him! The big boss was swearing at him like a trooper, telling him that he was not going to take away my Christmas

THE DARK SIDE OF THE BBC. A DISTURBING TRUE STORY.

bonus and that "Tony is alright". Eventually, after he had left the head chef grovelling for mercy in a heap on the floor, he turned to me and said "there you go Tony, you just take this cheque to Barclays bank on park lane and they will cash it for you". Then he shook my hand warmly and said "have a nice Christmas". I remember the pay cheque was for over £800 so I certainly did have a nice Christmas! Cheers boss!

I started to sleep with some of Patrick's models, most of whom were amateurs and completely unknown, but one of the girls was George Best's ex girlfriend who was doing some modelling as a new venture. We shared a spliff and had some sexy chat but I didn't manage to get any further...basically because Patrick was there with us. I didn't even manage to get her phone number. Maybe next time eh!

Well, for the first few weeks that I was in London Patrick was keeping himself occupied waiting in hotel restaurants so I spent a lot of time in the flat on my own, usually playing music. One day I was sitting around on my own when the phone rang: It was Beccy. Beautiful Beccy, whose picture had guided me through my last weeks in prison and helped shape my new life. Although I had asked Patrick about her - obviously she was one of the first things on my mind when I got out - he had just replied that they had fallen out and so I left it at that. Speaking to her on the phone, I was ready to steam straight into her, but there was a situation: She and Patrick had fallen out over some pictures, which she described as 'tacky'. He had so far refused to hand them over to her for disposal, claiming that he, as the 'artist' had ownership rights. To make things worse, she also claimed he had made some inappropriate comments about the Star of David she was wearing around her neck. He'd

THE DARK SIDE OF THE BBC. A DISTURBING TRUE STORY.

apparently told her to take it off as it portrayed a certain religious belief. Well, I hate to see a damsel in distress so I dug around the flat, found the offending pictures eventually (personally I don't remember thinking they were tacky, but these things are subjective I suppose) and posted them back to her. That one phone call was the first and last contact I had with Beccy. She didn't even phone back to thank me...although maybe my proximity to Patrick had something to do with this. I told Patrick about what had taken place but I don't think he was too impressed, I think he felt like he had lost face somehow in the stand-off and that I was the cause of it. Then we met Kirsten and Henna. Patrick had dialled a wrong number trying to call one of his mates and started chatting up this Danish au pair who answered the phone. Give him his due, Patrick can speak to the women and we starting seeing this pair of Danish Beauties regularly. Kirsten was the best looking so I ended up with her while Patrick had Henna. I seduced Kirsten within a couple of weeks, but I don't think Patrick got it on with Henna. I was getting on well with Kirsten, and I was enjoying having a fairly steady relationship – but my brother was starting to feel jealous, and it became clearer over time that he wasn't comfortable with my happiness. He'd decided he was going to drive us apart. When Kirsten and I did eventually have a big row and split up, Patrick soon moved in for the kill, but he was cunning: he played the sensitive man, told her he was going to cheer her up, offering to draw up a portfolio for her. Every girl loves to be made to look fantastic by a professional, and Patrick knew exactly how to play on that. He worked on her slowly. First he took pictures of her fully clothed, then topless, then naked. Within a few weeks they were bonking. I saw her as a bit of a bimbo...lovely to love, but I wouldn't have married her so it didn't particularly

THE DARK SIDE OF THE BBC. A DISTURBING TRUE STORY.

bother me. For Patrick however, she was his star prize. He fell deeply in love with her. I think he was so into her that he had to poison her against me so that he would be sure she'd never go back with me. Over time she started to show signs of disrespect, and I'd just laugh, because my feelings for her were all in the past and if she was going to be moody towards me, I didn't really care. He brought her up to see our parents in Birmingham. Like I said, she was his angel but eventually she left him to study politics at university. But this disloyalty was just the start. Never averse to talking about me behind my back, Patrick had started to put a negative slant on nearly everything he said regarding me. My Southern fame quickly turned to notoriety, and this notoriety was brought to the attention of the biggest scandalmongers of all. The British Broadcasting Corporation, who were based down the road in shepherds Bush.

It was the BBC that then eventually moved in on me, their people creating a conspiracy all around me, the new kid on the block, in order to scandalise me for their own enjoyment. They would all get involved eventually, from the lower ranks that pounded the streets to the higher echelons of power within the BBC and they began to use some highly illegal, technically advanced broadcasting methods in order to alienate me from society then eventually they would try to destroy my will to fight them back. They would use telepathy against me.

I was with Patrick one day towards the end of summer. The year was '87. We were walking down King St in Hammersmith when I stopped these two women in their late 30's and started to chat them up. I was only about twenty seven-twenty eight at the time, but I was always in the party, party mood. The blonde one of the two gave me her phone number and I phoned her up a couple of

THE DARK SIDE OF THE BBC. A DISTURBING TRUE STORY.

days later. In the meantime, word was beginning to reach me of Patrick's backstabbing. One of his female friends told me that he, in her words, "doesn't speak very highly of you". She said it as a warning I guess, but I didn't make the connection, and I remained gloriously unaware of what was going on behind the scenes. Anyway, this blonde woman invited me round hers to have a drink together, just the two of us. When I got there (courtesy of my trusty racing bike), she was excited, and made no attempt to hide it.

We went to her local pub which was somewhere just past Richmond, west London and we stayed there all night until closing time which was 10:30pm in those days. It turned out to be a good nights revelling actually, but I did get the feeling that quite a lot of people were coming into the pub and joining our company over a short period of time but they weren't really in the party mood like the rest of us. They just seemed to be watching me. I even began to suspect that they had followed me.

As it was, I said my farewells to the locals at the end of the night and left with this blonde. I still cannot think of her name! She took me to an off-licence she managed. It was closed, being near midnight by now, but she had keys. We acquired some booze and then it was back to hers where we partied the night away. I got *very* drunk, but I stayed alert, knowing I would have to cycle home. I remember this blonde getting her tits out for me and dancing round the room, but at some point a teenage kid kept on coming into the room then leaving again. He was passing messages to her son who was in his late teens and he was hiding in his bedroom.

I remember getting my bike from the hall in the small hours to head home, and I was confronted by the son. He stopped me and started acting as if he was

THE DARK SIDE OF THE BBC. A DISTURBING TRUE STORY.

some big time gangster or something, imposing himself on me to such an extent that I felt threatened. I gave it to him once across the forehead with the bicycle chain, then stood there for a second laughing at the three of them as they hurled abuse at me. I even whacked the blonde across the hand as she grabbed onto the wheel of my bike in an attempt to stop me from leaving. Soon I was heading back to Patrick's, but where the fuck that was I had no idea. The whole situation to me seemed hilarious, and I was still giggling when, passing an open field and in pitch-darkness, I was stopped by the police. A panda car had been following me from a distance for some time and there were two riot vans and two cars waiting for me by the field. They started to question me. I knew I was going to get a beating for what I'd done on my first time on their patch, but when they asked me why I did it I bellowed out "Well, he deserved it!" Said with a wicked smile on my face and still laughing. As soon as I said it, the atmosphere changed, but not in the way I was expecting. It was as if someone had shouted "we like him!". They seemed to show respect for me. I was confused, but the alcohol blurred my mind. I was taken to a police station in the South West and was charged with assault of some kind, but all the time they were buzzing around me, watching me, analysing me. I didn't know what was going on, but I felt like I was somehow special, like I was really famous. I had to appear at the magistrates court a couple of times after that, and I was stone cold sober when the magistrate eventually told me that the charges were being dropped, but the way he said it made me feel exactly the way I had in the Police station: like I was somehow famous. It was like he was amused that the police would do such a thing. But they had, and I was free to go. After that, I began to see and hear signs constantly, little things, hints that

THE DARK SIDE OF THE BBC. A DISTURBING TRUE STORY.

told me I was somehow famous. The most blatant was when Patrick and I walked into the local grocery. The shopkeeper's face lit up, and he called out "It's the famous brothers!"

I got arrested soon after for possession of cannabis; it was a joke really, I was carrying about enough for one spliff. But I was held in Ladbroke Grove station for a few hours, half-cut and loving it. I decided to play the police up, so I started wailing and howling like a cat before a fight. Within five minutes I was charged and released on police bail. They were so nervous of my new fame and notoriety that they just wanted me out of their way before I really started to play them up. I never even went back. I started to go up Ladbroke Grove, Notting Hill quite regular at this time to score weed. Usually I was the only white guy in the Blues, except when I was with Patrick, but I didn't mind. Patrick only came to observe the reaction I got from the black underworld. The fame was present even there. Although I was trying to keep myself on the straight and narrow, I still liked to smoke the 'wacky baccy' at this time, but I would eventually cut out that little vice in the near future. I was getting the feeling I wasn't too welcome in Patrick's flat anymore, and had to bail out rather quick like. I had hardly any money, but I landed a job at a Greasy Spoon café in east Acton. Although I'd gone to college after I left school, in order to learn my trade as a chef, I hadn't stayed there for the exams. The gaffer gave me the start anyway, and it turned out I was a class above. I moved into a B&B down East Acton but I only lasted there a couple of weeks due to my boss not giving me my wages on time. I had no choice really, but to go back to Patrick's. Things were a bit more chilled after our break from each other, and I wasn't unhappy there. But then we didn't really have a chance to get on each

THE DARK SIDE OF THE BBC. A DISTURBING TRUE STORY.

other's nerves because it was then that I met Judith, a black girl who lived in Blenheim Crescent, off Ladbroke Grove, Notting Hill.

THE DARK SIDE OF THE BBC. A DISTURBING TRUE STORY.

JUDITH'S FLAT.

I was drinking in the Coleville arms, which is on All Saints road Notting Hill. I got chatting to a woman called Judith and she mentioned she had a spare room in her flat. I knew she had her eye on me, but I wasn't interested really as she was too big for me. I took the room anyway of course, and quite nice it was too but Judith's cooking of Caribbean food was even better. I managed to get a job behind the bar at the Hombre nightclub off Oxford Street. I was only there for a week as a casual barman before I was sacked. Judith was still after me while I was living there, and when I slept with a girl there that I pulled in Hombre Nightclub she was put out, but didn't say anything. To be honest though, I had other things on my mind: like the night I'd been sacked from the club, and the more I thought about it in the following days, the more I began to question what had really happened. I never had a proper explanation from the gaffer, he mentioned something about me getting too much in the way of tips and something about it stuck in my throat. Everyone else seemed to like me, so why did this one guy have it in for me? I honestly began to suspect that I was the victim of some kind of conspiracy. If only I knew! It was about then that Judith moved in another guy, a black alcoholic guy. Judith was on some sort of medication herself for a psychological problem, nothing major and it never showed overtly. But this guy she moved in took an instant dislike to me. They shared the same bed, but I never heard any passionate noises so I don't know whether they were bonkers. Maybe he saw me as a threat - which I

THE DARK SIDE OF THE BBC. A DISTURBING TRUE STORY.

wasn't – but we definitely never hit it off. Anyway, I wasn't enjoying sitting around at Judith's much, so I made moves to find another job. Pretty soon I landed a spot valeting cars at Kenning car hire, which was on Holland Park road, Notting Hill. I didn't have a driving licence but that didn't seem to be a big problem. I was told I could have the job as long as I didn't drive the cars on the streets from the front of the garage where they are returned by customers to the back of the garage where they are brought in for servicing and valeting. But as soon as the gaffer was gone out for an hour or two I was off and away in these luxury cars, Lexus, Mercedes, BMW's etc., for a quick spin down to Patrick's just to show off basically. But the thing was, he was never in when I called to show off my prize possessions, so I was a bit deflated nearly every time. Until the one time I looked up at his window on the top floor and just saw him duck back out of sight when he realised that I was looking. That's when I realised that my own brother was trying to avoid me for some reason so I thought fuck him, he won't get any favours from me anymore. Ironically, it was on this occasion that when I returned to the garage, my gaffer was standing there waiting for me with someone I now know to be from the BBC. I'd been grassed up by the BBC to my gaffer!

My gaffer went ballistic at me, which was a bit of an overreaction really (he was only a little chap as well) but he was trying to impress this other little guy from the BBC I suppose. I had to just stand there and take it like a man. There was nothing I could say or do in my defence. I was caught red-handed! Thanks Beeb. Needless to say I was eventually sacked from there after only four weeks but the pay packet I got was substantial. For some reason my gaffer had paid me for four full weeks plus a big massive bonus just like the Charlie

THE DARK SIDE OF THE BBC. A DISTURBING TRUE STORY.

Chester Casino had. Except I reckon my gaffer only paid me the bonus so that I wouldn't seek revenge on him for sacking me that way at any time. I stayed away from Patrick for a couple of weeks after that but being as I was out of work for now I decided to pay him a visit one evening. I arrived at the house where his flat was on the top floor but he was not answering the front door still. I stood on the pavement and started to call his name out to him hoping he would answer me. He didn't. Instead the yuppie/tough guy who lived in the flat underneath Patrick came out and he was looking for a fight! He advanced on me so rapidly and with such menace about him that I suddenly changed from Mr. Cool into Mr. Angry. Now I am very experienced in the fight scenario so I knew that this asshole wasn't going to get his punches in first. I took one step forward to meet him and let off one of the biggest right hooks I'd ever given. It caught him full impact on his left jawbone and sent him crashing into the brick wall behind him where he lay motionless for a couple of seconds until his senses returned. I could have brutalised him in that state if I had wanted to but I'm not that way inclined particularly so I allowed him to get to his feet to continue the fight if he wanted to. He chose not to. Clever lad!

I had to rant and rave at him for a minute or so just to impress upon him the error of his ways then I told him in not so many words to stay out of it. I left still fuming at him but calmed down a short while afterwards.

    To my amazement, I walked into the cafe at the Last Chance Centre in Shepherds Bush after using the gym there two days later and lo and behold there was this same lad there working behind the counter. He quite sheepishly said "hello Tony" and now feeling a bit sorry for him I replied "hello" back to him. That was the last time I ever saw him again, but I did see his wife or

THE DARK SIDE OF THE BBC. A DISTURBING TRUE STORY.

girlfriend that he lived with the next day and the look she gave me as she passed me in the street could only be described as 'the look of amazement'. I mean her eyes were lit up like beacons as she bore her look right into me with a knowing grin on her face that stretched from ear to ear. This kid was starstruck! Not just so much for what I had done but for who I was. I thought she was just happy because I had given her bullying husband or boyfriend a lesson in life as I used to hear them rowing underneath us when I was staying at Patrick's. I was to see this look on people's faces time and time again in the forthcoming years, on both men and women so I began to realise that there was some kind of 'weird' conspiracy taking place and that a lot of people were loving me because of it now.

Further comments were made in my presence which fuelled my belief in a 'weird' conspiracy, whereby people seemed to know and say things about me even when they didn't know me, such as the time I went up the West End on my own for a few drinks and I heard a girls voice in the crowded pub exclaim aloud, "there's that boxer". A verbal 'exclamation' I now know which was designed to help me by keeping the 'hyenas' – the people who now followed me around to cause trouble for me - at bay.

I was still staying at Judith's at this time when one night I went outside the flat, just to check the streets for some moody reason, whereupon I saw a group of those yuppie student types that like squatting in empty houses and using alcohol and drugs. I stood there observing them for a few seconds when I heard a girl's worried voice say "see, he does come out!.'" Hmmm! I thought. Then they all disappeared into their squat which was about three or

THE DARK SIDE OF THE BBC. A DISTURBING TRUE STORY.

four doors away. This was at about 1 o'clock in the morning. The plot thickened.

A few nights later, one of Judith's female friends arrived at the flat with one of her male lovers. His name was Marcelle, a likeable but dodgy character who seemed to be well known and friendly with the whores in Soho. She was bisexual and also had a female lover who worked as a prostitute from her flat in Earls Court. There was also a big black guy with them. As it happened, we three lads ended up chatting outside to the student type squatters that lived up the road. They were sitting on the steps outside their front door and we were standing. It was a mixed company, both boys and girls that were squatting. The conversat I let him give it the large for a minute or two, then I just punched him one right in the mouth. He fled the scene with his tail between his legs muttering some obscenities to himself but the girls were not very happy with me either. I told them I didn't give a fuck and that I was a Birmingham fighter and we do not stand for that kind of thing up there!

They were all silenced and left the scene. We three lads were returning to Judith's flat when the big black guy turned to Marcelle and exclaimed, "he split his lip". So now the word on the street was, "Tony split his lip".

Then there was Frank. Frank was a cockney, and then some. Now frank actually gave me a car – it was only a shitty old escort, but he gave me this car anyway, and it felt a bit weird to be driving again but I was ferrying Frank around a lot of the time so I picked it all up again pretty quick. One time, I was driving with Frank and the black alcoholic guy who was staying at Judith's, and we were driving slowly in a traffic jam which was making me a bit vexed so I starting revving the engine like a formula one racing car and instantly the

THE DARK SIDE OF THE BBC. A DISTURBING TRUE STORY.

queue behind me disappeared. Then the whole street disappeared of cars in front of me for as far as I could see.

I was a bit freaked out by that, how a whole road full of cars could disappear into thin air just because I showed a little bit of bravado.

At that time, Frank had some Polish folk renting a room off him and along with Frank they conspired to commit the next act of violence against me. The Polish folk told Frank that I had stolen a gold necklace or something like that from their room and then proceeded to turn him violently against me, which wasn't difficult because Frank secretly wanted to bash me up anyway so that he then would be the talk of the town. It all came to a head about a week later when Frank landed at Judith's flat one day with a smile on his face and invited me round to his house for a few drinks. Of course I accepted and went along with him and one of his short-arse cronies whom I had never met before. The Polish folk were upstairs in their room when we arrived and I led the way through the front door and into the living-room. Within a split second and before I even had a chance to turn around, Frank came at me from behind with a pickaxe handle. He put everything he had into his swing and the wood landed its best shot somewhere just above my right ear. To be honest though, I didn't even feel it really. I only realised what had truly happened when I saw Frank standing there with the pickaxe handle in his grasp. Well, someone should have informed the little idiot that I am not the sort of person that likes to be attacked from behind with a pickaxe handle. The rage that engulfed me was overwhelming. I unleashed a barrage of head shots at him that sent him stumbling backwards across the room where he ended up on his back sprawled across the settee. It was game over for him.

THE DARK SIDE OF THE BBC. A DISTURBING TRUE STORY.

I even throttled him as he lay there, shocked and wounded on the settee, for at least half a minute but I knew that I had terminated his attack so I began to ease off a little on him until I eventually ended my counter attack. His short-arsed crony had by now skulked to the relative safety of an armchair away from the war zone, so I didn't even bother with him. He wasn't in my league anyway. The Polish folk then made an appearance in the hallway, but a very brief appearance. They listened to me shouting at Frank for wrongly accusing me of stealing from him but when I walked into the hallway to check them out; they quickly fled out of the house before I found out that it was them that were the original instigators. After that, things were not the same between me and Frank. I began to drift away from him over time until I only saw him when he was of some use to me.

One of the people I'd met while living at Patrick's was a woman called Annie, a blonde woman who I'd met while she was out walking her dog. She had a daughter, Angie, who was seeing a black guy called Trevor. I'd stayed in contact with them, and Trevor's friendship came in real handy at this point: One of his mates, Christine, told him about the place next door to her house in Cathnor road, Shepherds Bush – the previous occupier had passed away and it was just sitting there empty. Trevor was a big black guy who was into bodybuilding and Christine was a white woman from Manchester. Trevor and I decided to enter the building via a window and then squat the whole house which was divided up into three flats. I was worried someone might hear us smashing the window, but Trevor came prepared with an industrial roll of Sellotape. It was a bit grim to be walking around this dead bloke's home, even

THE DARK SIDE OF THE BBC. A DISTURBING TRUE STORY.

though none of his stuff was there. Thank God the electricity was still on though.

FIRST SQUAT.

When I got the money owed to me from Kenning car hire a couple of days later I did some serious shopping for paint, wallpaper and brand new furniture, until eventually I ended up with a pretty decent place around me. I left Judith's

THE DARK SIDE OF THE BBC. A DISTURBING TRUE STORY.

place on good terms and I moved into my first squat on Cathnor road. I had the top floor flat, Trevor and Angela (Annie's daughter) had the middle floor flat while the basement flat was left empty. We were all living next door to Christine. I would later discover that people used to call her 'the mad woman'. Christine was an ex-prostitute and an interesting woman. I thought at the time that she'd taken a shine to me because whenever Trevor said anything to me she thought was out of line; she'd jump to my defence. In retrospect, I think that maybe she'd somehow picked up on the fact that there was something weird going on around me, and against me and wanted to try and defend me from it. And there definitely was something weird going on around me, only I didn't have a clue what it was as yet. I'd begun to notice people's reactions to me when I was out around the city. It was almost as if they could read my mind; as if they were intuiting how I was feeling. Sometimes they made it obvious, actually saying things to me about my state of mind, but mostly it was more subliminal. Things were picking up speed.

It was the end of 1987 when I had an experience that, although insignificant at the time, proved to be the key to my understanding of the extent of my situation. I was out drinking in White City with some of my workmates, in a boozer used by a lot of actors and media darlings. In order to relieve myself I went into the gents, and spotted Mel Smith the BBC actor who was just coming out. I smiled at him and said 'alright Mel', and the glare he gave me stopped me dead in my tracks. I don't know whether my attempt to communicate with the mighty Mel had so angered him that he used his position to turn the BBC against me - certainly until then their interest in me seemed to be occasionally benevolent – or whether they had already made their decision

THE DARK SIDE OF THE BBC. A DISTURBING TRUE STORY.

and mighty Mel was simply acting on that in his disrespect towards me. Whatever the case was, their latent malice came to fruition in 1988.

At the beginning of '88 I managed to get a maintenance job in Next, the big retail store on Kensington High Street. By this time Christine also had turned against me as a result of my spurning her advances over Christmas. I stayed a month at Next, and got on well with everyone there because they seemed to enjoy my funny sense of humour and often made it obvious to me that I was somehow famous. There was just one incident that made me feel uncomfortable: As I walked out onto the shop-floor on what turned out to be my last day there, the security guy there was talking to someone I didn't recognise at all but was now obviously from the BBC. I didn't intend to eavesdrop, but it was pretty hard to not see him point me out to the stranger. Later that day, when I found a watch on the counter in the empty staff canteen, I picked it up thinking I would find out who it belonged to and return it to them myself. It was obviously a girl's watch so I thought I would be the hero when I returned it. But I didn't have a chance. I had to leave the watch on my desk in my office for some time as I was called away on a job and upon my return found the watch was gone. I was soon confronted by the Next security guy who dangled the watch in front of me telling me it was found in my office and that I had stolen it. He took me to see the manageress where I was promptly dismissed. It was clear that I had been framed. I tried to tell her the truth, but she had none of it. The BBC had sunk their fangs deep into her as well.

Christine had introduced me to one of her friends, a girl with a funny leg. Her name was Laura. We hung out with her quite a bit, and one

THE DARK SIDE OF THE BBC. A DISTURBING TRUE STORY.

time when we went round to her place just off the Goldhawk Road, Freddy Knowles was there. Now Freddy Knowles was an ex middle-weight boxer and had supposedly travelled the world on the boxing circuit. He was still giving it large, despite it being common knowledge that he was a has-been. He'd recently found himself homeless and was kipping at Laura's place……. although she was clearly unhappy with this arrangement and wasted no time in offloading him onto me. I wasn't impressed, but agreed that he could stay at my squat if he paid me fifteen quid a week. Soon after that, I started seeing an Irish girl who I bumped into while making a late-night visit to the 'offy. We were in my room several nights later lying naked on the bed after just having made love when Freddy burst in, yelling accusations that I'd used his toothpaste. He was serious aggro and after several seconds of slandering me he charged across the room to me and whacked me in the face as I lay on the bed and then ran out of the room. I wasn't going to take that. I grabbed the table leg from under my bed and, still stark bollock naked, chased after him and whacked him one hard enough to stop him in his tracks. I'd split his head wide open just under the right ear and he began cowering from me, pleading "Tony stop! You've done me, you've done me!". Being of an honourable disposition, I didn't inflict any more serious damage on him and told him to stay the fuck out of my room. It was soon after that little incident that we received a summons to appear in Court, due to the fact that the housing association that owned the property wanted their house back. Trevor, Angela and myself duly attended Court on the said day some weeks later and pleaded our case to the presiding Magistrate. He, in turn, seemed to be smiling in recognition of me and I was happy when he ordered a stay of execution of our eviction and gave us three

THE DARK SIDE OF THE BBC. A DISTURBING TRUE STORY.

months to find alternative accommodation. That was plenty of time for me to be drawn into the next phase of my stay in London.

THE DARK SIDE OF THE BBC. A DISTURBING TRUE STORY.

SECOND SQUAT.

Anyway soon after that I moved out of that squat and into another, above a girl called Debbie. I'd met Debbie at Christine's flat and we'd had a kiss and a cuddle, but nothing had happened between us. She wanted me though, and had offered me this squat above her place just so that she could have me. Unfortunately this drove a wedge still further between me and Christine and she began to turn against me even more. We still saw each other and went out drinking together occasionally though but she now always had it in her mind to see my downfall. There was one time we'd gone to the local pub when we noticed Gary McMillan, who had just done ten years for burglary – although Christine reckoned there was more to it than that. She'd heard that, or so she told me, not content with robbing her possessions, Gary decided he wanted some extracurriculars from the female occupier into the deal and had touched the girl's breast he was burgling. I didn't believe any of it though but that's what she told me. In retrospect, it seems clear that he had been brought down to the pub by a girl from the BBC to check me out. He turned away after we acknowledged each other for the very first time and subtle as fuck, he mumbled

THE DARK SIDE OF THE BBC. A DISTURBING TRUE STORY.

"as long as he doesn't want to fight", obviously telling the BBC that he wasn't willing to go toe to toe with me. Safe, I thought. Unfortunately, his girlfriend who came into the pub long after Gary had gone decided she was going to shit-stir. Christine pointed her out to me in the same pub that night, after I was full of the holy falling down water and told me to ask her for myself if Gary was a 'nonce' as she called him . I wouldn't believe it myself so I did ask her. I said to her "Gary's not a nonce is he bab?" She didn't reply though, she didn't tell me he wasn't a nonce or that he was, she just got up from her table and left the pub like it was on fire! I was a bit confused but still didn't believe it anyway. I finished the night's revelling, getting pretty drunk in the process, and headed back to my second squat above Debbie's. Unbeknown to me, Gary's girlfriend must have gone straight to him after she had left the pub and told him that I had called him a nonce. Which was not so.

But she must have given him that impression because when I got to the squat, Gary was waiting outside for me, probably in that same BBC girl's car whom Christine later described tactically as "one of his 'friends'. He was brandishing a knife as he approached me in the hallway outside Debbie's flat and I spun round to face him. "What have you been saying about me?" he demanded. I was very co

    A week later, after a bit of a session with Christine involving Vodka and beer, I bumped into Gary in a bar up the top of Cathnor Road. We spotted each other and – to my great surprise – he walked up to me, acting all matey and gave me a big hug and called my name out loud to the whole pub like I was his long lost friend or something. I had to smile. Despite the fact that his brother was there with him, he was bricking it, and trying desperately to prevent any kind of a showdown...although his reticence could have been simply due to the

THE DARK SIDE OF THE BBC. A DISTURBING TRUE STORY.

fact that he was unarmed, because when he let me go and I jokingly asked whether he wanted it with the knife "now or later", he had to smash his glass in order to threaten me. But this was done after I had turned my back to him in order to get a drink from the bar. My recollection is a little woozy after that (probably due to the extent of my inebriation) but apparently a couple of our mates had been standing in the background and stepped into the fray. After a bit of scuffling, Gerard (who was married to Theresa) got the glass stuck in his face by Gary. Next thing the whole pub erupted and steamed into Gary and his brother before throwing them both outside. The Police were called and I kept asking Gerard, shall we go? Shall we go? as I had a stolen cheque book in my pocket that I was holding for a youngster called Mark Cheeseman (I was using them as well of course as I now felt that my courage was being tested by London). When the Police arrived they ended up arresting Gary and his brother but also they arrested my man Gerard who was by now covered in blood from a deep gash in his head. Of course, at this point I only knew a little of the BBC's involvement in my fate: This was made clearer to me during the next chapter of my life. I must stress however that I did realise that most of the BBC womenfolk seemed to hold me in high esteem. They saw the way I was always smiling, even when I had half of London out to scandalise me led on in their efforts by a majority of the men folk of the BBC.

THE DARK SIDE OF THE BBC. A DISTURBING TRUE STORY.

JACKIE'S FLAT.

Not long after the Gary McMillan incident I moved into a flat in White City, owned by some girl I'd met at a party, but I didn't last there long. Her name was Jackie and she was a small blonde with two sisters called Kim and Maxine who were also blondes and they all had nice legs. Jackie had a daughter of

THE DARK SIDE OF THE BBC. A DISTURBING TRUE STORY.

about four years old called Kimberly who was a beautiful kid and was definitely going to turn the boy's heads' when she hit sixteen. Jackie was a new tenant to this flat and had no furniture when she moved in, so when I joined her I let her have all of mine from my second squat in Cathnor road. I gave her brand new fitted carpets, new double bed, television and hi-fi etc. I even stayed in some nights to baby-sit Kimberly while Jackie went out for a drink with her friends but she kept coming home pissed every time and I just had enough of it.

After a while I was introduced to her sister Maxine's boyfriend whose name was Dave. We used to call him 'Dave the rave' because he was a bit of a div really although he was a six footer. One day, me and Dave went for a job interview on the other side of London with 'Newsfleet' a big trucking company. The jobs on offer were truck drivers and I had a bent driving licence given to me by Frank so that I could get a job. Conspiracy or what, but one of the employers came out of the office with some lorry keys and asked Dave to park this lorry up that was in need of repair. He didn't ask me to do it for some reason, the real star of the show. Dave bottled it though so he asked me to park it up, which I did. I really thought I had impressed the employers enough with my driving skills for them to give me a job but after keeping us waiting for over an hour, one of them came out and said "sorry, we have no work for you". I was gutted and angry at the same time, but then I realised that I still had the keys to the lorry in my pocket because I was so excited that I had forgotten to return them. I felt I had to strike back at them for disrespecting us and wasting our time. We decided to 'borrow' the lorry to get us home to White City that night as we had very little money left so we hid ourselves and waited for the cover of darkness which was only about an hour away. When it was dark, we

THE DARK SIDE OF THE BBC. A DISTURBING TRUE STORY.

jumped into the truck and sped off down the dirt track to the main road, disappearing into the congested traffic. We were howling and laughing most of the way back home because we had taught the bastards not to take the piss! I let Dave have a drive but he smashed into a parked car somewhere in White City so I had to jump out of the cab, run round to the driver's side and take over again. I dunno, You just can't get the staff can you?

After only a few weeks my little blonde lover turned against me because of what was manifesting around me and when we had a big row she called the police on me. She just wanted to rob me of all my furniture before I left basically but I had to bail out before the Police arrived. I wasn't welcome there anymore.

THE DARK SIDE OF THE BBC. A DISTURBING TRUE STORY.

THE BROWNHEADS' FLAT.

I then found myself in a White City high-rise, in a ground floor flat with some 'brown heads' (heroin addicts) I'd vaguely met a couple of times. It wasn't my scene, but I was homeless so I had to stay there and watch them stick needles into their arms every other day. It was horrible and degrading staying there so at about six o'clock one morning after having spent the night drinking, I returned to Jackie's flat on the other side of the White City Estate to see if we could make up. She was no stunner or anything really but I just had to get away from the brown heads. She lived on the ground floor so I climbed over

THE DARK SIDE OF THE BBC. A DISTURBING TRUE STORY.

the balcony and looked into the bedroom window knowing she would still be asleep. To my horror I saw Dave the rave lying in bed next to her and I went ape-shit. I hammered on the window until they both woke up, cursing Dave all the time in the process. Then Jackie did something really weird. Instead of bolting and latching all the doors and windows as you would expect, she walked to the front door and opened it wide then walked back into the living room. The little femme fatale only wanted to see the great Brummy fighter in action didn't she? I had to oblige so I ran round the outside of the building to the front door and entered the flat to find them both waiting in the living room. He didn't look too confident but I didn't care, he had offended me so he had to pay the penalty. She just stood there with a wicked grin on her face. Was she somehow told to open the front door? Anyway I wasn't reasoning with the situation. I told Dave what a wanker he was and that it was time for fisticuffs! I then stepped into range and let off a big right hook that connected square on his jawbone with some force. His legs buckled under him as his jawbone fractured on the left side but he didn't go down. This just made me even angrier but I could see that his lack of confidence had now turned to that of immense fear so I gave him the chance to take a 'dive'. I waited for a few seconds then I hollered at him "look at him, he still hasn't gone down!"

When he realised what I was saying to him a few seconds later he went down into the chair next to him rather sharpish and stayed there motionless and speechless. He must have thought he was dealing with some kind of mug I guess but who could have put that thought into his tiny mind? I didn't have time to speculate because in all the commotion Jackie had sent her little daughter out to her sisters to phone the Police so I had to move my ass again.

THE DARK SIDE OF THE BBC. A DISTURBING TRUE STORY.

It was only a matter of minutes but the Police had surrounded the perimeter and I had to bluff my way out of the proximity past them. It was unnatural how they could have deployed so quickly, it was as if they were telepathically aware of the situation (and how!). I had no choice but to walk straight past them as their radios shattered the early morning silence. I was stopped by a Policeman about 200 yards away who asked me my name. I gave him a false name but I was so cool, calm and collected that he actually believed me. He knew I was his man really but my sheer confidence threw him and he let me pass after I even had the nerve to tell him that I had seen this Tony fella going the opposite way. I managed to get about 500 yards away from him before they landed on me in numbers. I was duly collared and taken away in handcuffs to Shepherds Bush Police Station. I was charged with grievous bodily harm (later dropped to actual bodily harm) but while I was in custody, Jackie told the Police that I had assaulted her as well and she even told them about the lorry I had taken from Newsfleet. She mentioned nothing about Dave's involvement. I was hit with a few charges and bailed to appear at West London Magistrates Court in the near future. I had to be escorted by Police back to her flat to collect my clothes that she had so far refused to give me but she wouldn't give me my furniture. I told her she could keep them.

So it was back with the brown heads again as I was still homeless, but the next day I saw the BBC begin their first blatant 'interviews' of people, right in front of my very eyes. These tossers were trying to provoke people to be violent towards me it was obvious, but for how long now? The onus was on the fact that I had given Dave a hiding and what was going to be done about it.

THE DARK SIDE OF THE BBC. A DISTURBING TRUE STORY.

When I walked past this little 'interview' scenario, I heard the guy being interviewed declare: "well, he deserved it!"

He was saying that Dave the rave deserved what he got and that he (the interviewee) was not going to mess with me. HA! HA! What a blow to what was now becoming a 'criminal conspiracy' being orchestrated by the BBC. But it wasn't all over yet!

The next day it was the turn of Jackie's big brother to get in on the act. Firstly though it was Dave's mom who sent her boyfriend to capture me and bring me back for questioning. He was a biggish sort of guy, a black guy who I had met before along with Dave's mom and they both liked me. I went with him to see Dave's mom who was waiting outside the brown heads' squat where I was staying. I just told her the truth about what had happened previously, how I had caught Dave in bed etc. and she believed me and didn't want to take the matter any further. Then Jackie's big brother arrived on the scene with his other sister Kim in tow (conspiracy or what?). He started to give it the big 'un' straight away saying that I had assaulted his sister etc. Then he attacked me. There were a few people watching as they saw me swing this guy round as he attacked me and smashed him up against a parked car. I grabbed him by the hair and punched him repeatedly in the face. When I did stop, Dave's mom's boyfriend stepped in between us and said "all right, that's enough!" It wasn't the fact that he was protecting big brother so much, it was because he wanted the fight to end with me as the winner! Which it did... So the underdog strikes again!

Now I really was becoming famous amidst this 'weird' conspiracy and funnily enough I was beginning to revel in it, becoming more and more drawn

THE DARK SIDE OF THE BBC. A DISTURBING TRUE STORY.

into it as each day passed. Unfortunately for me though, I had now become the centre of attention wherever I went and was attracting all the unsavoury characters, the low-life's and the criminals onto the scene who wanted to test me and compete against me in order to see me fall from grace. I was slowly being drawn back into a life of crime which I had so wanted to escape from but because I now continually achieved more fame and respect by passing all tests and confrontations, I began to enjoy my role as the underdog, thriving in what was potentially a sordid situation. But I was still living with the brown heads and when my chance came to escape I shifted out rather speedily into a flat two storeys above them.

THE DARK SIDE OF THE BBC. A DISTURBING TRUE STORY.

THIRD SQUAT.

The flat itself was owned by one of Christine's friends who gave me the keys because she was moving somewhere else and she told me to squat the place if I wanted to. So I certainly did!

It was within days of moving into this my third squat that I realised the radio was making comments on how my state of mind was. It was a girls voice that began to make comments on how relaxed I was feeling, even going as far as to comment on the relaxing decor of my surroundings as if she could actually see them. Basically, this girl was emphasising my cool and calm nature at this time because of the mass hysteria that was engulfing the BBC conspiracy since I took on and destroyed Dave the rave at six o'clock in the morning and got arrested by Police for it, followed by my demolition of Jackie's big brother soon after. No-one had come to take me on for daring to commit this show of force so it now appeared obvious that they were all too nervous of me, and this girl knew it. She also knew that the BBC were coming out in droves onto the streets now to investigate the catastrophe and that some of them would be seeking revenge. I knew I was being scrutinised somehow but I believed the BBC were

THE DARK SIDE OF THE BBC. A DISTURBING TRUE STORY.

using hidden cameras to monitor me at this time. Little did I know that they had projected what were probably microwaves onto my person and that these microwaves were connected to the electrical activity of my brain. This enabled them to monitor my thoughts, emotions and even my sight which they could then broadcast into the minds of anyone and everyone else within a large radius around me. They could even broadcast their own 'voices' into people's minds, thus enabling them to instruct people to say or do things that were detrimental to my well being.

That was only the tip of the iceberg. As it happened, I continued to search for work in order to maintain a decent standard of living and the sixth job I landed in London was another one I got from the job centre in Shepherds Bush. I started working as a fitters mate for a company called Jeff Air who were based somewhere just outside London. This firm specialised in installing ventilation systems into big shops and offices and were apparently quite good at it as well. I was given the start by the site foreman who interviewed me although he did seem to be a bit wary of me, as though he was aware of something about me which made him inquisitive. There were three main contracts in operation at the time I started work installing ventilation systems, one was on a big department store on Kensington High Street in Kensington, another was on a building site in Notting Hill Gate and the third was on the construction of ...... Wait for it ...... The new BBC studios in White City. I worked on all three sites during my stint with Jeff Air, but after only a couple of weeks I soon got the feeling of the conspiracy going on around me. Eventually, one of the foremen came to test me out by taking me to the BBC site in order to get me hulking tons and tons of aluminium shafts around, on my own. I worked all day doing

THE DARK SIDE OF THE BBC. A DISTURBING TRUE STORY.

that until the foreman had had enough of it and let me go home early. I worked hard to impress that day but it was all in vain because soon after I was laid off. Now I was being persecuted by the BBC bad guys but I never let it bother me too much.

I soon began to notice a show of force by the BBC whenever I was out walking the streets in the locality. The most prolific time was the very first time when they made it blatantly obvious. It was supposed to frighten me. I left the new flat I was squatting in White City and made my way up Loftus Road towards Shepherds Bush. This is a dead end road so there is never any traffic and it is always quiet. As I made my way along this road happily whistling to myself as I approached another road that cuts in from the right, which is also another quiet road, they appeared in force from out of the blue. There were about a dozen cars in all in a single line convoy that drove past me, not at speed but all of them revving up their engines as they passed and all of them looking straight at me in defiance. I tell you, if looks could kill! Most of them were only female chancers that obviously gloried in the fact that they could abuse the powers bestowed on them when they joined the BBC. None of them got out of their cars to confront me so I just thought that typical. I was now beginning to see both sides of the coin.

The good guys and the bad guys I called them. I saw some of the good guys a few days later in a pub on Uxbridge Road, Shepherds Bush. It was one girl really, but she was flanked by two of her male colleagues from the BBC. She had come to this pub, which was my local at this time, specifically to see me in the aftermath of the Dave the rave fight. She knew I had sent the BBC paranoid since then and she knew I was becoming popular even in the wake of

THE DARK SIDE OF THE BBC. A DISTURBING TRUE STORY.

all the telepathy that was being used against me, but was I deserving? Was I bigheaded? Did I have the brains to go with the brawn? She had come to investigate. I arrived on the scene and I could tell straight away what was in motion. I got an awful buzz off it! She was a very classy lady, probably not quite thirty years old but she probably held a high ranking position within the BBC. She made my eyes shine and caused a smile to light up my face. I knew she was not a threat to me and maybe even liked me by the way she studied me from her table in the centre of the room. Her two male colleagues seemed pleasant enough as well. No-one spoke whilst they were there, they all seemed to be smiling as well at the enormity of the situation and after about five minutes, our guests from the BBC got up and departed, leaving their unfinished drinks behind. They had seen enough to be convinced that I should have friends in high places and would use their influence to convince others as well.

I hooked up again with a nipper called Mark Cheeseman, who I'd met the year before when I was squatting above Debbie's. The police used to call him Burglar Bill - he was into kiting, stolen chequebooks and stuff, although as the nickname implies his main hobby was burglary. We started drinking together, then going out stealing; why I was getting involved in this kind of life again I know was not what I wanted at all but I felt like I was being tested, against which I was obliged to prove myself or lose face in the midst of a huge conspiracy. This is backed up by the fact that to begin with, the police didn't hassle us at all even though they knew exactly what was going on. But suddenly they seemed to change their minds, and they were all over us, arresting us for the slightest misdemeanour. They didn't like the fact that I was an ex-con becoming big time in London so they continued to follow me around

THE DARK SIDE OF THE BBC. A DISTURBING TRUE STORY.

in order to get me arrested any way they could. But not all of them did this I must stress. Only the bad guys.

I was walking out of the station fresh from being charged and bailed one day when my attention was drawn to a white van driving past me as I walked up the road. It was a BBC van, logo-ed and blatant, and the guys in there were openly filming me, one of them was hanging out of the side door with a television camera on a mount pointing straight at me, his face was lit up with a wide grin and his eyes shining like beacons. It was then that the past three years began to make more sense. I walked home in a daze...could the BBC really be engineering these situations around me? What was their motive? Why had they chosen me? I had no answer to these questions at this time that were spinning through my mind, but at least now I knew that there was a reason for the events that surrounded me. Mark and I went on a bit of a spree after we were arrested for the first time. The police had never particularly scared me but, fuelled by my knowledge that events were being orchestrated around me anyway, I stopped caring what happened to me at all and sank deeper into a life of crime and violence. So much so that I ended up getting arrested again a few times until I finally ended up in court facing quite a few charges ranging from one of stealing a ham sandwich to one of violence. I was set up by the Police and the BBC on a few occasions I know now. One time was when the Police evicted an American girl called 'Skate' and her boyfriend from their squat near me on the White City Estate. They landed at mine for a while but I didn't like them after a time when I saw them injecting heroin into their arms one day in my bathroom. It was sickening! There was blood squirting from their arms up the walls and ceiling. Anyway, after a couple of

THE DARK SIDE OF THE BBC. A DISTURBING TRUE STORY.

weeks I was set-up again on a charge of burglary. Skate's boyfriend had been telling me that some guy he knew had stolen a cheque book and card from him and wouldn't return it. He talked me into going with him as a bit of muscle so that he could get it back. When we arrived at this guy's flat, he wasn't in (wouldn't you know) so Skate's boyfriend decided to burgle him and rob his video. He smashed the window in the front door and disappeared into the flat, returning a few seconds later with the video. I stayed outside as I didn't want anything to do with burgling a dwelling. As soon as the video was outside the premises a policeman arrived on the scene appearing from around the corner of a road which was about fifty yards away. Skate's boyfriend made a run for it with the video before dumping it on the pavement and absconding. I decided that I didn't want to get myself arrested for someone else's crimes and ran off in the opposite direction before jumping a garden wall and hiding under a hedge. There was police everywhere within seconds and I was soon detected in my hiding place whereupon I was handcuffed and taken to Shepherds Bush Police Station yet again. Skate's boyfriend was already there when I arrived and I went ape-shit at him for getting me arrested. I tried to tell the Police that I had nothing to do with the burglary but they insisted that I was acting as lookout and I was formally charged and held in police custody overnight. I appeared at West London Magistrates Court the following morning before Mr. Goldbottom (I think that's how the name is pronounced) where I opted to be tried at the Crown Court. Obviously aware of all the conspiracy (and telepathy) around me at this time, Mr. Goldbottom smiled at me and said "yes, I think that is the best bet". Initially I wasn't given bail, my previous record meaning that I was going to spend the pre-trial period in Wormwood Scrubs. Admittedly I

THE DARK SIDE OF THE BBC. A DISTURBING TRUE STORY.

acted a bit psycho in there...so I wasn't surprised when after one week I got bailed and released. I had to get my uncle Paddy to act as surety for me in the sum of five hundred pounds. I returned to the squat in White City, probably just to piss the BBC off. Mark and I started going out robbing from cars again, but this came to an abrupt end when Mark started hearing voices in his head (telepathically from the BBC studios) telling him to test me to the extreme. In the end he went berserk after a drinking session and started smashing up cars and robbing from them as we were walking back. The police landed almost instantly - there was a guy standing up the end of the street watching as Mark lost the plot, this guy must have been part of the set-up, there to tell the cops when to come and get us. It had all been so weird for a long time now but I ended up in West London Magistrates Court again and this time there was no chance of getting bail.

THE PRISON CIRCUIT.

So I landed in Brixton prison with no hope of bail, losing my beloved squat in the process plus I lost all my property I'd left in the squat. They shifted me around a lot before the trial, and I did brief stints at all the main London jails

THE DARK SIDE OF THE BBC. A DISTURBING TRUE STORY.

starting off with a month in Brixton, a month in Pentonville, a month in a remand centre out in the sticks close to Heathrow Airport, then when there was a period when all the London Prisons were full to capacity I did a few weeks being moved around all the police stations throughout London until I finally ended up doing the last six weeks before my trial in Wands worth Prison. When I found myself in Wands worth I somehow got honoured with the main job in the kitchens: I was basically the cook, in charge of the ovens while everyone else ponced around washing up and the like. This wasn't a particularly unhappy period for me: although everyone there knew all about me and everyone was on my case to such an extent that one day, a girl's voice on the radio said to me, "it is better to fight and lose than not to fight at all". That had a major influence on me.

The BBC were broadcasting to them all in the Prison, influencing their actions subtly in an attempt to break me. They were using their secret advanced broadcasting method, whereby they broadcast directly into people's minds telling them what to do and say that would be detrimental to myself. But mostly, they broadcast my vision to all around me so that everyone would know exactly where I was and what I was doing and could be intercepted in any given place. I was a human camera at this stage but I would later learn that their capabilities were far, far greater than just that.

Happily my job in the kitchen did have a few small perks...I had easy access to the luxury items that were brought in and these could be sold off for smokes. Unfortunately the one time I managed to lay my hands on some Jaffa Cakes - a scarce commodity on the inside - the con' I passed them on to tried to mess with me and sold them off himself. So when another brummie, a black guy told

THE DARK SIDE OF THE BBC. A DISTURBING TRUE STORY.

me what he'd done we decided to take the matter up - us brummies stick together. We landed on him in his cell and my brummie mate whacked him proper in the face, smashed him up good he did leaving this guy pissing out blood from his mouth. The guy never even retaliated. After that, all the smokes he was owed from the sale of the Jaffa cakes came straight to me, the rightful recipient. Of course I sorted out my brummie mate who had told me what was going on as well.

The white guy in the cell opposite me was also a fighter, although he was only a kid, and he was always getting into scraps with black guys. Once, he decided to mess with my cell mate, an African called Timidayo Ashinai. They'd headed down to the pots and pans area, a nice little place in the kitchens where all the fights take place. There was a huge crowd down there and as soon as things kicked off it seemed that my cell mate was overpowering this kid so his cell mate, a black guy, jumped in on the action and started kicking my cell mate who was also a black guy. Well, I wasn't standing for that so I jumped in as well. I grabbed this interfering black guy and spun him around to face me, then I gave him such a pisser in the mouth that he went flying backwards for what seemed like an eternity, before crashing into the wall that housed all the pots and pans which came crashing down around him. The sound effects were excellent as you can imagine and this con' just stood there dazed and mortified. I could see that I had knocked the fight out of him with my big right hook so then I roared "just stay out of it" to him and anyone else that had the same idea. No-one moved for the rest of the fight and my cell mate went on to victory. My same brummie mate who helped me out earlier with the jaffa-cakes said that I'd "done us brummies proud", despite the fact that he was a black

THE DARK SIDE OF THE BBC. A DISTURBING TRUE STORY.

guy himself. The fight scenario kind of fizzled out after that; the kid probably thought he had no chance against the two of us. But the next day he walked up to me in the exercise yard when I was on my own and told me I'd made a big mistake. He reckoned I should have stuck up for him as my cell mate was only "a fucking African" but I told him in not so many words that this fucking African was my cell mate and cell mates stick together. I had a wicked, confident grin on my face as I spoke, just to let him know that I would take his head off if he started, so he didn't. He sauntered off with his interfering cell mate into the background and I just laughed out loud to myself. Then in the next second, one of the Prison Governors, a well dressed woman, came out onto the exercise yard to view me. She wore that same wicked, confident grin on her face that I had on mine so I knew there was a link of some kind in operation and that she was buzzing off me. I also knew then that there was a fight of some kind going on against me and that I appeared to be winning.

I had been involved in seven fights in London at this time and a large number of confrontations and I had won all of them, but apparently, the BBC still continued to ridicule me and invite people to take the piss out of me. They would have to be taught a lesson I giggled to myself! I sort of had an idea that they were up to something a bit strange but I still had no idea what was really going on.

When I moved from the kitchens to a hospital cleaner's job, I still saw the young white kid around and the buzz was 'Does Tony want to fight?'. He'd been fired from the kitchens even before I left - the screws waited until I was there and fired him in front of me. I saw him around looking aggro at me but I

THE DARK SIDE OF THE BBC. A DISTURBING TRUE STORY.

wasn't going to go out of my way to fuck him up (I'd already done that) and he soon disappeared.

Admittedly the BBC were still quite operational at this point. Another blatant intrusion into my privacy was a repeat of the incident in the squat when they spoke to me through the radio. This was before I got moved to my new job as a cleaner in the hospital (I was told it was a lot quieter there and the pay was better). I'd been having a smoke of black in my cell with the other kitchen workers; we were just chilling out and chatting but we were all really stoned when suddenly everyone but me dived into bed and under the sheets. I didn't know what was going on but then the radio started talking to me. We were listening to Capital Gold, our usual station, and the presenter jokingly said 'Bang your head against the wall Tony', and I did so, laughing all the time because it was so funny, so mad!. Like I said everyone in there knew about the BBC, and were proper paranoid of me by now. This was particularly evident in the sports sessions, where most people couldn't even keep their minds in line long enough to remember how many laps they'd done whilst circuit training. Of course I always finished first anyway; I was dedicated to my training, and put all my effort into it. Even the gym screws used to watch me and count the laps I did to make sure I wasn't cheating in order to finish first. I finished first every time because I was fitter and faster than any-one else and no other reason.

Eventually I went to Knightsbridge Crown Court to be sentenced. My Barrister was arguing for twelve months which would mean I could walk then and there as I'd already served half of my sentence, but the judge upped it to fifteen months so I was back in Wands worth for another four months because in

THE DARK SIDE OF THE BBC. A DISTURBING TRUE STORY.

those days a prisoner had to serve two thirds of his sentence if it was over twelve months. Thank you your honour!

This passed quite uneventfully though considering the circumstances, except that once more Capital Gold spoke to me through the radio: The presenter I remember as Andy Peebles said "Hello Tony" as I rested my head down on the pillow and relaxed next to the radio and I just went "Hiya" back to him as though it was a completely normal thing to do. This made him recoil back away from his microphone in shock and amazement at how completely normal I was amidst all the growing use of telepathy. But nothing else of note happened inside Wands worth until the day before I was released when I was taken down to the bath house to have a shower. That was a bit strange in itself because it was not shower day for our wing in the prison. When I entered the bath house, I knew straight away that I had been set up because the whole place was full of convicts milling around whereas there should only be one landing allowed in at any one time for security reasons. I was supposed to have felt threatened I guess but I never did, I just got on with my own business and if anyone wanted to do something about it they could just do it. No-one did though because although I didn't know it at the time, my sheer confidence was being broadcasted to all around me and it was them that felt threatened, not me. Eventually, after the screws let the situation go on and on forever, thinking that there would soon be a confrontation, one of the con's came up to me and asked me how my sentence had gone. He was checking me out! I'd done it real easy to be honest, but not wanting to sound big-headed or anything I replied "well, I'm a little battered and a little bruised, but I done it!"

THE DARK SIDE OF THE BBC. A DISTURBING TRUE STORY.

To this response, my little friend immediately exclaimed aloud, "Tony's alright!" Suddenly everyone started to react amiably as one to my little friend's comment so I knew there was something a lot more quicker than the 'grapevine' in operation and the more and more of these little scenes I saw over the years, the more I knew what was really happening. Anyway, the whole of the bath house was eventually cleared of us con's and everyone went back to their normal routine. With the big conspiracy ended, nothing else of note happened before my release from Wandsworth Prison except when the girl on the radio, again on Capital Gold said "...and he goes back to his roots". I knew then that she meant I should return to my hometown, Birmingham. But there was no way I was going back without showing my face in Shepherds Bush again first. I had to let them know I'd done it. It was just before Christmas 1989 when I got out. 8 am I walked free, but even before I left the prison I was getting hassled by some kid trying to sell me a pair of trainers in the reception area. I took them off him and trapped, thinking nothing more of it. It didn't occur to me that he might be insinuating something about my fitness even though I had run them all into the ground for the last six months. Outside the gate there were two guys sitting in a white Transit Van and I heard one of them mumble "he hasn't learnt his lesson yet".

This was of course just a made-up bullshit comment but this guy was out to cause trouble for me from the relative safety of his dilapidated van. If I'd known straight away that he was talking about me I could have confronted him about it, but the BBC had been laying quite low during my final weeks in Wandsworth so they weren't at the forefront of my mind. The conspiracy only came to my awareness again later on that day when I met Frank again in a bar

THE DARK SIDE OF THE BBC. A DISTURBING TRUE STORY.

back in Shepherds Bush and, right in front of me, he turned to someone else and said "well, he's gotta learn, hasn't he?". This, combined with the guys in the van and the kid with the trainers made me realise that there definitely was something extraordinary going on again. It was now that things hit the big time.

I stayed at Frank's for a little while. I was using him, and I was unhappy about the situation but it was only a stop-gap measure. I needed to get back to Birmingham now as I had lost my beautiful squat in White City. The council had moved some new tenants into the premises during my long absence. I got a ticket to Liverpool instead of Birmingham from the prison on my release in order to throw the BBC off the chase, then jumped a train to New Street, Birmingham. I wasn't planning on there being an inspector on the train going home to Birmingham so when I saw a guy in a British Rail uniform approaching, I realised that I would have to do some fast talking as to why I had a ticket to Liverpool when I was on a Birmingham bound train. I was just about to deliver my blag when I realised that the two guys now sitting opposite me with the big bright shiny eyes and demonic grins staring at me, were from the BBC and that they were letting me know that I was being followed home. I felt confused and angry at the situation but the inspector stopped me from doing anything rash by intervening between us. I spent the rest of the journey home seething to myself and planning how to escape from the BBC who had somehow managed to home in on my whereabouts. Where the hell were their cameras? How were they keeping tabs on me?

I got off the train at New Street pretty sharpish and ducked in and out of the crowd as I made my way to the turnstiles, losing my persuers in the process.

THE DARK SIDE OF THE BBC. A DISTURBING TRUE STORY.

I'd made it! I handed my ticket to the Indian guard on the turnstiles and was mortified and confused again when he looked straight at me. His eyes lit up like beacons as if he knew who I was and he bore that same demonic grin on his face that told me he was also aware of the weird conspiracy that had taken place for the last few years. It also told me that it was not solely constrained to London and that it was now still around me, in Birmingham.

THE DARK SIDE OF THE BBC. A DISTURBING TRUE STORY.

THE DARK SIDE OF THE BBC. A DISTURBING TRUE STORY.

BIRMINGHAM.

THE DARK SIDE OF THE BBC. A DISTURBING TRUE STORY.

THE DARK SIDE OF THE BBC. A DISTURBING TRUE STORY.

THE FAMILY HOME.

I landed back in Charles Road, the house I'd grown up in. My parents had split up while I was in London and it was just my dad there, Patrick Hickey Snr. with my nephew Shaun Fleming when I arrived. I didn't want to involve any of my

THE DARK SIDE OF THE BBC. A DISTURBING TRUE STORY.

loved ones in the shit that was happening to me so I kept quiet, managed to chill out a bit and just acted normal with them for a while. What I didn't know was that the BBC was already way ahead of me. At one point Shaun turned away from me when we were talking in the bedroom and quietly muttered the word "No". It didn't really mean anything to me at the time, but in retrospect it seems clear that the BBC had asked him telepathically if I still wanted to fight them. I guess he didn't want the BBC to think that I would cause them any aggro' and so they would leave me alone, but these people get their kicks out of persecuting others this way which is why they built and developed this technology.

My dad, Patrick Hickey senior also knew about the BBC and because he was paranoid about the situation he eventually tried to get the doctors onto me, the victim, rather than stand up against the BBC. I suppose it's because my Dad isn't a fighter like me, he's a graduate with a couple of Bachelor Of Arts degrees from Limerick university, so I've no bad feelings for that I guess. When the doctor turned up though, you could tell that he didn't want to mess with someone like me and he left without much hassle. The Doctor that came to my house to see me was our family doctor from 'The Park Medical Centre' on Tennyson Rd, Small Heath. His name was Dr. Harrison and he told me to see a certain Dr. Nath, a psychiatrist in Small Heath. I did this a few days later like a fool and thus began the dirtiest cover-up to hit the streets of Birmingham. It began with Dr. Nath who immediately opened a file on me after I opened my heart to him about my problem with the BBC and then it slowly spread throughout the medical profession. I was immediately diagnosed as paranoid schizophrenic in a bid to cover up for the illegal activities of the BBC. So I

THE DARK SIDE OF THE BBC. A DISTURBING TRUE STORY.

fucked Dr. Nath off because I had a vague idea of what was going on now and discharged myself from the service of this criminal conspiracy. Unfortunately, these people now had their foot in the door and began to plot against me in the background, waiting to pounce. It was good to be back home though and I felt comfortable there, but soon I needed something to occupy my free time. I started training again as usual , and I credit this with doctor Harrison's decision to stay away from me: I've not always been a beefy geezer but when I started building up my muscle and my skills with the punch- bag, Dr. Harrison decided there was no way he was going to risk involving himself in a criminal conspiracy against me. I was too big now for him to officially put his neck on the line in a cover-up as he could see I was not an easy push over and would not go down without a fight. So, their initial plan scuppered, the BBC also resorted to their most effective medium for controlling people's lives: television. More and more at this point I felt both the TV and the radio talking to me as if they knew exactly what I was thinking.

One time I was watching the TV and I glanced out of the window to see two Asian neighbours outside, mimicking talking into microphones. I knew they were observing me and mimicking reporting back to their masters, the Beeb. So I went up real close to the TV set and growled menacingly at the presenter – the presenter stepped back looking shocked and said out loud, "His will to fight on is as strong as ever!" Then when I went out onto the streets, I first saw scotch Maggie, Sadie's sister who's eyes were shining as she smiled at me in recognition of the fact that she heard that my will to fight on was as strong as ever. Telepathy, through me, was now to become an everyday event in Birmingham! So was the conspiracy for or against me that went with it! The

THE DARK SIDE OF THE BBC. A DISTURBING TRUE STORY.

Police were just as bad. They would encourage people to fight me in the streets, often arranging fighters to appear in front of me as I walked alone to wherever I was going. One time I was up town with some old mates I hadn't seen for years. One of them was called Richard, a blonde haired guy who used to drink with us up town in all the best pubs, such as 'The Parisienne' and 'Cagneys'. I left them at the end of the night to get the last bus home from town from Colmore Row. I noticed two policemen hiding behind the beef-burger stand and knew straight away that something was amiss. Then this big tough guy just walked over to me out of the blue and picked a fight. I was taken by surprise as this guy suddenly threw a punch that landed above my left eye. He was wearing a ring on his finger and this cut my skin as it landed causing me to bleed. Two seconds later I had knocked him from one side of the road to the other. Then and only then did the Police step in to stop it. One of the Police started creeping to me then by saying things such as "and I know you could take me on if you wanted to Tony" and so on, but I was still raging and refused to calm down so they took me to a little police station just around the corner for half an hour until I did calm down.

When I left the station I had missed the last bus home so I had to walk home from town. As I walked down the

CCoventry Road just outside town, I was met up by some stranger who proceeded to tell me that the guy I had the fight with was from Coventry and that he was supposed to be a hard man but he was just a big poser. I hadn't even mentioned anything about a fight to this stranger! Another time the BBC arranged a fight for me in the Porsche nightclub on Regents Park Road in Small Heath. I was drinking on my own in there at the time and was feeling a little

THE DARK SIDE OF THE BBC. A DISTURBING TRUE STORY.

drunk towards the end of the night after having had an excellent time. I made my way to the toilets outside the dance halls to have a piss when I saw this guy there looking a bit dodgy. His name was Liam Harkin and he was supposed to be an Irish boxer from Small Heath. He was being egged on by a BBC girl standing next to him outside the toilets. They had been waiting for me to arrive as this guy was supposedly going to do me in. I didn't even realise what was going to happen as I was in high spirits from the drink and the good time that went with it, so I just passed them by and went into the toilets. I did my business in the latrine and proceeded to zip up my flies when he pounced on me from behind. Real brave like!

He threw a flurry of punches at me from behind and by the time I realised what was happening, I was on top of him on the floor having overpowered him and was just positioning myself to punch his lights out when some black guy behind me grabbed both my arms and pulled me off him, allowing him to rise up. Next thing I knew he had bolted out of the door and disappeared. I started to laugh out loud to myself at the situation as I splashed water over my face, when this same black guy said aloud "OK, that's enough now". I don't know why I didn't steam him for pulling me off but I didn't. I was still laughing when I got outside onto the street heading home when I saw that the street was full of BBC girls milling about looking a bit para'. I heard one of them say "he enjoyed it" to which another girl said "alright, that's enough now". That just added to my amusement as I made my way home on foot.

      I started going out drinking with a good friend I'd hooked up with again. I had known Dessie Mulcahy most of my adult life, and he often jokingly referred to himself as my bodyguard, although we both looked out for each

THE DARK SIDE OF THE BBC. A DISTURBING TRUE STORY.

other. Whenever we were out on a drinking session around Birmingham, I knew people were aware of me and passers-by would make comments relating to my predicament. During one of my drinking sessions I saw John Pertwee, another BBC stooge who played Dr. Who in the popular BBC television series. I could tell he was aware of me but he was careful to avoid any kind of acknowledgement. He just stared moodily at the ground as I walked by. He was accompanied by this young Asian girl who smiled all the time at the situation.

At this point I still had not confided in anyone about the BBC's operations. To be honest, I was still having difficulty believing it myself, even though all the signs were there. And they were getting more blatant all the time. A girl on the TV said to me "they've got microphones so powerful they can actually pick up thought". Until this was said, I knew that somehow people around me were picking up on how I was feeling but I had no idea that they could actually read my mind. But I was picking it up slowly. I went to the Jobcentre in Small Heath one day looking for some kind of work. It was not long after Christmas in 1990. I applied to an advert for someone to help run a market stall in Aberdeen, Scotland. I thought this would be a good opportunity to escape the melee that had followed me from London. I would see if it would follow me to Scotland. An interview was arranged for the next day at a house on Coventry road, Small Heath, which I duly attended. The employer was an Indian guy who had a beautiful wife and two teenage sons who would all, except the wife, work the markets in Aberdeen. The two sons made me giggle with laughter because they both were dressed in turban headwear but spoke with a Scottish accent. I got the job with them and started work a few days later. My employer and I would take it in turn to drive the Mercedes van

THE DARK SIDE OF THE BBC. A DISTURBING TRUE STORY.

through the night until we arrived at the market place in Aberdeen. It was about a twelve hour drive because we couldn't drive too fast due to the loaded van we were laden with. When we arrived at the market place, we had to unload the van and set up stall straight away as we usually arrived at about six in the morning, then we had to work the stall all day until the end of trading for the day. We normally worked from 6pm Thursday to 6pm Monday and I was paid £90 for it. My employer had apparently just sold off three shops he owned in the area before moving to Birmingham, but had kept the market stall and a three bedroom flat in Aberdeen. When we finshed work for the day, I would drive one Mercedes van back to the flat and my employer would drive the other one which was kept permanently in Aberdeen. We would all eat a big curry that his wife had prepared for us in Birmingham then they would spend the evening relaxing whilst I would go out sightseeing around the town or sometimes visit the local pub on my own. It was whilst I was out on my own that I began to realise something was amiss. The conspiracy had followed me! People were making comments about me and about what I was thinking at the time. The only difference was that these people were Scottish, they were from Scotland, so they hadn't followed me from Birmingham, I had somehow brought it to Scotland! I spent the next few days in confusion, wondering how the BBC could be doing it until I finally came to the conclusion that they were watching my movements by some kind of very powerful camera from a satellite or a series of satellites or watching me through the television. I would persist with this theory for a good few months before I learnt the truth. Anyway, the weeks flew by as I commuted between Birmingham and Scotland, knowing all the time that there was no designated free zone for me. The conspiracy in Scotland became more

THE DARK SIDE OF THE BBC. A DISTURBING TRUE STORY.

and more open as the weeks went by, as the inhabitants got more and more used to the existence and misuse of telepathy. I was a marked man. I found the Scottish to be a very patriotic bunch who didn't really like foreigners on their patch. They didn't seem to mind me though because I'm Irish born and from Irish decent and because I used to take the piss out of them and their accent. But they knew I was supposed to be the victim of a BBC controlled Telepathic conspiracy, and sometimes made it obvious. One day, the conspiracy did rear its ugly head in the market place where I worked. I had been doing the Scotland run for about four weeks by now and I could see with my own eyes that some of the BBC had followed me to Scotland, with their big smiling faces and mad, shiny eyes. They were also the only ones with English accents. Again, it was the women that made the positive comments mostly while some of the men were always negative. I was working on the stall with my boss and his two young sons when one of the other Scottish market traders came to our stall and picked a fight with the younger of the two sons. The market trader was about 40 years old and the youngest son was about 15 years old. How brave of him!

I know now that it was all for my observance, but at the time I got the feeling that this trader was letting me know that Indian people were not welcome on their patch, so I watched them argue a little until I saw this guy move forward and slap the young son in the face. It was uncalled for and pretty much a cowardly act so I stepped in between them and pushed Scotty back away from the son and said "No! leave him alone". I was prepared to take this guy out and he knew it so he backed away and returned to his stall. My boss, the Dad, was furious but because he was only a little bloke, he couldn't do anything himself about it. He ranted and raved for a few minutes until the Aberdeen Police

THE DARK SIDE OF THE BBC. A DISTURBING TRUE STORY.

arrived and arrested the whole Indian family and took the three of them away to the Police station, although they were the victims! I told you the Scottish were very 'patriotic'. I was left on my own to look after the stall by myself but I didn't even really know the prices of any of the clothes that were being sold. I was thinking this to myself after they had all gone, when, wouldn't you know it, it seemed like the whole of the market place shoppers descended onto my stall and started trying to buy up everything. They were obviously out to do my head in but I was cool under pressure. At first I was running around the stall just guessing the prices as I didn't have a bloody clue but when the full stampede arrived, I just started selling everything for a fiver. They were all getting a bargain and I was in control again because I had a fixed price in my head. I even started playing up to them all by acting like a real experienced market trader, calling out to them, "roll up, roll up, pick anything you like ladies, everything a fiver". I sold more in the space of ten minutes at that time than we all did for the whole day. When my employer finally returned with his two sons in tow, he was quite pleased to see the wad of money I presented to him. The whole Scottish thing fizzled out after that because of the aggravation and eventually my employer told me he would sell off the two Mercedes vans and the market stall itself. He wanted me to go as a witness at court for him over the aggravation and I said I would at first but that fizzled out of my head as well eventually when I returned to Birmingham. He was claiming racial discrimination but I knew he had no chance of winning because as I said, the Scotch are just too 'patriotic'. I got a phone call from the English Police about four weeks after I stopped working for the Indians. Apparently, one of their Mercedes vans had been set on fire and the insurance claimed on it but a note

THE DARK SIDE OF THE BBC. A DISTURBING TRUE STORY.

was found in the cab with my name, Tony Hickey and my phone number on it. They said it looked like whoever set the van on fire was trying to make it look as though I had done it. I told them I didn't do it and it was left at that. That's how I was repaid for not going as a witness. Say no more! I spent much of my spare time doing hard, physical exercise which I continued with even when I landed my next job in Birmingham.

At the time I was feeling quite content with my life: I started working as a fitter-engineer for Initial UK in Acocks Green in April of 1990, a company that makes public toilet hand-dryers and hygiene products. At the first interview however, the BBC made a blatant appearance whilst I was waiting outside the interview room, waiting to be seen. It was a youngish lad of about 25years old who made his presence known to me. He just came in and stood in front of me with those mad, shiny eyes and big grin for about 5 seconds, then disappeared. I felt a bit vexed after he'd gone because I thought he could have been there to ruin my chances of a job. He personally didn't, but all that would come later. I passed the first interview which was with the foreman but then I had to have a second interview one week later with the works manager, Geoff Perkins, who also wanted to see me. I obliged of course and at the second interview he asked me about my previous employers. When I told him I had just spent the last four years in London, he replied "but it didn't quite work out huh?" I could have told him I kicked ass down there but I just continued to smile sweetly. He was making it obvious that he was aware of the big conspiracy around me. He gave me the job anyway, though probably because he thought of me as a bit of a novelty, someone to take the piss out of. Initial UK are also a very large laundry service. My dad wasn't around much, so when

THE DARK SIDE OF THE BBC. A DISTURBING TRUE STORY.

I wasn't working I could chill at home on my own. I bought a punch bag so I could keep on training and stay in shape, and my original plan was to set it up in the garage at Initial, but when the gaffer heard about it he suddenly became very moody with me. This was about five months after I started working there. He didn't want me becoming too big in the eyes of the staff at his firm so I was sacked not long after for bad time-keeping, although clearly that wasn't the real reason. I'd been given a verbally flexible start time of 6:30-7:00 am, and usually came in at 6:45. I was later told my start time was 6:30 and so I was always coming in late for work. All the engineers told me that "Perkins is out to sack you", so I knew it was because I was using the punch bag there. I lost interest in working for him then. When he eventually did sack me, he had one of his charge hands present with him in the office. Perkins told me he was there as a witness, but really he was there in case I gave Perkins a punch in the mouth. (He'd had one recently from one of the other workers he'd wanted to sack). I just told him to shove his job where the sun doesn't shine! Because I was sacked from my job, the Social suspended my benefits when I eventually did have to sign on. I took my appeal to a tribunal, with the help of the Citizen's Advice Bureaux, which resulted in the reinstatement of my benefits, although not before a little mudslinging between my ex-boss Geoff Perkins and I had taken place. I got several hundred pounds in back benefits and the woman from the CAB Who had handled my case seemed to be over the moon that I had won my case because she obviously knew of the BBC Conspiracy against me and the control of people by telepathy that went with it. I was looking for work again but it didn't prove easy. Every interview I attended displayed the BBC's arrogance in thinking they could manipulate and control my life. Several

THE DARK SIDE OF THE BBC. A DISTURBING TRUE STORY.

times I was offered work at the initial job interview but later rejected. Always there would be someone present, even if only for a few seconds, who I would later realize was one on the Beeb's payroll. I gave up looking for work eventually as it was pointless trying whilst I still had the BBC on my case. Then I met Pauline, a girl from Small Heath who I dated for eighteen months on and off. She told me I was very passionate in bed amongst other things so I must have treated her well, as you do. I took her down to London for a couple of days to do some shopping on Oxford Street and we stayed at Patrick's flat overnight. We split up eventually when the BBC turned her against me. Around that time I started seeing a girl called Shabeena Naz, a pretty Asian girl who lived with her mum, sister and brother on flora rd. Yardley. I used to call her Tina for short as her pet English name. We were dating et al though we had to keep it quiet. Her family wouldn't have taken too kindly to her hanging around with someone like me whatever the context as I wasn't even Asian. She used to tell me she had the feeling she was being watched all the time and although I didn't say anything, I knew who was behind it. Eventually the BBC did manage to turn her against me, using their telepathy to influence her mind.

And still, whenever I went out people would make comments as they walked by. Some were critical, some complementary, some designed to weaken my morale, others to reassure me or show faith in my struggle. Finally I told Des about what was happening. He just laughed and changed the subject, a little too apprehensive at this time to involve himself fully in talking about my situation. With the money I had saved from Initial UK and my DSS settlement I decided to take a driving lesson one hour before my driving test and see if I could finally get a license. I wasn't surprised when I did pass, although my

THE DARK SIDE OF THE BBC. A DISTURBING TRUE STORY.

examiner seemed really unhappy with me after he had passed me. It's possible that he had been instructed by the good guys from the BBC to pass me and my skill made it impossible for him to fail me. I remember Des asking me what would I do if I failed my test, about one week before I took it and I gave off one of my mad laughs that made my eyes light up and shine as I answered "I'll just crash and burn and turn to shit". Then at that point a BBC Man appeared out of nowhere also with the big shiny eyes and smiling as if to let me know he liked me and was going to use his position to help me. Whatever the case, I passed and the car I bought came in real useful when my dad threw me out of the house for bringing girls back and basically, to rid himself of me and the plague that followed me around everywhere I went.

The plague of course being the BBC and their cursed telepathy.

THE DARK SIDE OF THE BBC. A DISTURBING TRUE STORY.

HOMELESS.

I spent a while kipping on the settee of a guy known as The Barclaycard Kid. He was about fifty years old. This guy was a friend of my mates Tony and Jamie Ash's and had earned the nickname because Tony and Jamie used to con him into using his bank card to withdraw money from the cash point to fund their drinking sessions. The guy was a bit of a nut by all accounts, on alcohol and anti-psychotics and who knows what else. I only stayed at his place for a few days. Tony and Jamie and some other guys came round one night while he was

THE DARK SIDE OF THE BBC. A DISTURBING TRUE STORY.

in bed and we started drinking and making a bit of a noise. The Barclay card kid burst into the room stark- bollock naked from his bed and started ranting and raving at all of us for taking him for a mug. Tony Ash just took one look at the idiot standing there in the altogether and told him to shut the fuck up basically. It was really funny but that was when I knew I had to get out. The guy was obviously off his head. I bought a car and spent a week sleeping in it, parked up in The Ackers Activity Centre in Small Heath. This wasn't as unpleasant as it sounds, and I had a few laughs, especially when Jamie and I brought two girls back there in the car. We stayed up talking all night, and at about 6am I went out to the little hut nearby for a walk with one of the girls and asked if we were getting laid that night – she just laughed and walked out of the hut somewhat amused but reluctant to lose her virginity on this occasion. I didn't pursue it any further when she told me of her virginal status as I'm an honourable bastard and she probably wasn't even sixteen anyway. Later that same day, I allowed one of the girls to have a quick drive in my car as I thought it was safe to do so because we were well away from any public road and it turned out to be a very quick drive at that. She drove the car straight into an iron bollard and writ off the whole of the left wing. I was so taken aback by this that I just burst into laughter at the sight of the wreckage, while Mammoth (another one of my nephew Shaun's mates who had joined us on this little escapade) and the rest of them looked on apprehensively at me, probably expecting me to go berserk. I wouldn't normally go berserk at any juvenile girl, so when she realised this, she just smiled herself and exclaimed out loud to her friend, "he just laughed!" This made me wonder later on if she was acting upon instructions to perform against me from the dreaded 'voices' in her head and I came to the assumption

THE DARK SIDE OF THE BBC. A DISTURBING TRUE STORY.

that she probably was. I dropped the two girls back home to their own back yard on the other side of Birmingham as a matter of urgency when I found out they were both a pair of runaways. The demolition driver was called Ann Gerity.

At the same time I had got myself involved in a non-dwelling burglary with my nephew Shaun Fleming and his mates, my first step onto the wrong side of the law since I left London. Again, I don't know why I got involved in this, as I know I didn't really want to at the time but I can only think that the BBC were influencing their actions by making them feel as though they needed to 'perform' in my presence and I felt obliged to take part as a spectator. I never played any part as I could tell they were all 'acting', mostly I just went along for the ride and watched as Shaun and his colleagues did the business. We were caught and I was given community service doing cleaning at a child's nursery in Yardley - the kind of single placement normally reserved for people who were trustworthy, though they only gave it to me because of all the Telepathy and they felt I needed to be kept isolated from the others. It was dull as ever really but occasionally Shabeena came up and we'd head out to the privacy of the shed. These trysts were essential in alleviating the boredom of the work, and eventually managed to get me out of the nursery altogether when Shabeena was discovered in the shed by the manageress - who'd locked the door after calling me out to carry on with my work and poor Shabeena had been locked in by herself, eventually banging and shouting for several minutes before she was heard. I got moved onto a gardening placement with everyone else for that and given an extra 40 hours community service as punishment. Anyway, much as I was enjoying slumming it, I needed to sort out something a little more permanent. I informed the council that I was homeless and they

THE DARK SIDE OF THE BBC. A DISTURBING TRUE STORY.

made an appointment for me to see the Housing Officer within a few long days. I sat in the waiting room some days later for several minutes listening to the conversations that were going on around me. It was eerily quiet really because they were all listening to me 'thought broadcasting', but I was so calm and confident that I was hardly even emitting any thoughts. The staff there must have liked that because suddenly I heard the Housing Officer say out loud "We're going to re-house him" and as I looked up, this girl, who definitely must have been from the BBC, stood up looking as para' as a parrot and skulked off out of the room and onto the streets.

MY OWN COUNCIL FLAT.

So against the BBC's wishes I got placed in a flat in Small Heath, and the council came top of my Christmas card list that year. It was early 1991. The place was beautiful, freshly cleaned and in a great location. Of course it wasn't

THE DARK SIDE OF THE BBC. A DISTURBING TRUE STORY.

furnished but my mate John Ellis sorted me out with a TV, fridge and chip fryer from a flat that someone he knew had vacated and I got a loan from the DSS for the other essentials, such as carpets, bed, three piece, cooker etc and I had a place of my own, in my own hometown! I actually moved into my new flat on Hob Moor rd in April 1991 and immediately began to settle in. Within a couple of months of living in my own flat and venturing into town most of the time to face the almighty BBC controlled conspiracy, I soon hit upon the idea of going to the BBC's arch enemies- ITV, to see if they would help me in my plight. I made my way down to Broad Street where Carlton, ITV's Midlands operators, have their headquarters. But I had no joy. I spoke through an intercom to someone in authority and told him of my plight, and then I asked him if there was anything they could do to help me. The guy seemed amused at my situation I could tell. He even had the nerve to laugh out loud. He just blatantly replied "no". I went berserk at the asshole, swearing like a trooper into the intercom before telling him to go fuck himself and his Television Company. I left Carlton Television studios in a bad mood that afternoon and made my way back into town. The telepathy was rife in Birmingham at this time with just about everyone being involved. I was about the only person that was not experiencing this highly intrusive phenomena but it was all directed against me via the general public. Walking through town was a bit of a task in itself with just about everyone I passed making comments about me or about what I was thinking at the time. These comments were usually restricted to one word or short phrases in order to keep people thinking it was all just a game and were nearly always disguised in a completely irrelevant sentence to confuse me but

THE DARK SIDE OF THE BBC. A DISTURBING TRUE STORY.

after a short time I began to recognise the continual use of the same words and phrases.

It was infuriating at times and often led to confrontations and the more I confronted these people that were abusive towards me, the more I was respected by the people of Birmingham and even the BBC themselves, who knew me from the beginning as a likeable rogue and were out to make sure I stayed that way, or fall heavily by the wayside. Of course I couldn't please everybody and so the abuse continued. I decided to try another avenue, and travelled down to the Evening Mail offices on Colmore Circus, the biggest of the local newspapers. I told the security guard there that I had a story. Boy, did I have a story. He phoned the big boys upstairs and one of the journalists made his way down to see me. He arrived in the reception area within minutes and I could tell from his approach from the distance that he was buzzing his tits off. This guy was going to kick BBC ass! He relaxed into the conversation as I told him all about the BBC and their illegal invasion of my privacy etc and he was keen enough to take on the story as a journalist. Keen enough that is, until two BBC operatives sauntered into the building looking stern, but clever enough not to look stern into my eyes. There was a male and a female together, both probably around thirty years old and they both just fixed their stern look onto the floor. My man from the Evening Mail slowly bottled it unfortunately, when he saw the two BBC stooges arrive on the scene, until eventually he told me I had to get evidence of hidden cameras and microphones etc. How was I supposed to get evidence of hidden cameras and microphones etc when I couldn't even see them? The guy was brushing me off now, unwilling to be seen making a stand against the almighty BBC conspiracy. I left the offices of

THE DARK SIDE OF THE BBC. A DISTURBING TRUE STORY.

the Evening Mail in a state of confusion that afternoon and made my way home to Small Heath. When I finally realised what had taken place, I was very angry and wished I'd have known straight away about the two BBC operatives, so that I could have confronted them there and then in the Evening Mail offices, thus letting my man know that I was a force to be reckoned with. Things could have been so different if I had, but it was too late now! The weeks passed by and I soon started to hear people proclaim that the BBC should "pay" for their invasion of my privacy and the word "money" was continuously mentioned on a daily basis, even by the Police. I only ever sought to gain my freedom and a return to a normal life but the BBC was having none of this. It was just a game to them, a wicked, cruel game with my sanity at stake!

Eventually I realised I had to take them to the Civil Court and this was confirmed when a BBC woman passing me in the street muttered "at last": at last I had realised what I had to do.

I travelled down to the Citizen's Advice Bureau in Birmingham City centre and as I passed a Dixons electrical store on Corporation street I heard the radio call me "The Six Million Dollar Man", which I knew was a message from one of the Beeb saying "Okay, we are allowing you to take us on and sue us". The receptionist at the Citizens Advice Bureau had a stern look on her face from the moment I walked in, obviously in an attempt not to show any paranoia. I told her the deal and I was left in the waiting room for a quarter of an hour while she made arrangements for a solicitor to see me. I used the time to relax psychologically but as my psyche spiralled down to a state of total relaxation I was interrupted by a strong discharge of energy from my head and I knew something was amiss straight away. I knew something negative had

THE DARK SIDE OF THE BBC. A DISTURBING TRUE STORY.

been broadcasted to everyone around me and this was soon confirmed when a female solicitor who was now standing in the middle of the room said "I can't do it". She had obviously been told by one of the bad guys from the Beeb not to take on my case now, as he didn't like it when my psyche went downwards, he was one of the shitheads that always tried to force it upwards towards a state of paranoia. I realised it was all over so I asked her "What am I supposed to do then?" and she said "I feel sorry for you". I stood up and left at that point, and was soon given a particularly harsh glare outside by a woman who must have been from the BBC. A few weeks later I wrote a letter to the editor of the Evening Mail. They at least had shown willingness to help until the BBC showed their faces. In it I asked them again for their help, describing what had happened at the Citizens Advice Bureau and how I had not been successful in my bid to sue the BBC, even though it was the BBC who had told me that that was what I must do. I then waited a couple of days just to make sure that the Evening Mail had received my letter and then I went into town to do some shopping, though really I just wanted to see what sort of a reaction I would get. While I was there someone from the Evening Mail just walked up to me and said "We'll get it for you". This made me think I was on the right track, but ten minutes later I saw a tall black guy in the Pallasades eyeing me up...I had an idea I was supposed to fight him but he was grinning like a Cheshire Cat. How could I know I was supposed to fight with someone smiling like that? He didn't make me feel threatened or anything. So I left him alone and continued with my window shopping, until I soon realised that an eerie silence had fallen over the town, no-one was saying anything to me as they passed by me in the street. I put two and two together and came to the conclusion that because I

THE DARK SIDE OF THE BBC. A DISTURBING TRUE STORY.

didn't fight the black guy in the Pallasades, I had been framed again and was probably being ridiculed to high heaven by the BBC. Then the same Evening Mail guy made an appearance again and he gave me a real nasty look as though he was scorning me for not fighting the black guy in the Pallasades. I didn't know what to think then, so I eventually went home, somewhat angry and confused, phoned the Evening Mail two days later to make sure they'd got my letter, and the editor just acted dumb, saying "What letter?" In a real sneering tone. I hung up and left it at that. My realisation that the Evening Mail weren't on my side now was soon justified a matter of weeks later. I'd been assuming that the BBC were like the gutter press and that the Evening Mail were somehow more respectable, but as I walked along Somerville Road in Small Heath I heard a woman's voice from one of the cars that passed me say "They're into the filth just as much as we are". This was a woman from the BBC telling me that the Evening Mail were no more trustworthy than them. I soon began to see for myself that she was telling the truth of course.

I was at home one evening watching the television when the BBC decided to scare the shit out of me with a demonstration of their power over me that their technology could inflict. I heard a man on the television say the words "and he sees the light and goes to meet his maker". I had no idea at the time that he was referring to me and I certainly had no idea what he meant but it all became clear after I went to bed later that night. I was in my bed just slipping into the realms of a deep sleep when I felt something around my neck like a rope or something. it was in fact a chain of electricity. It soon began to tighten and tighten until I was being strangled to a point where I began to scream out in panic as I genuinely thought they were trying to kill me. I knew

THE DARK SIDE OF THE BBC. A DISTURBING TRUE STORY.

straight away it was the BBC and I began to scream out "I'm normal, I'm normal", trying not to appear frightened. I was standing up against them even in my sleep. Then I could feel my mind rising and rising until eventually everything went white inside my head as though I had died and gone to meet my maker. The strangulation stopped and I met my maker in the form of a procession of BBC women who all appeared to be climbing down from the heavens on a ladder. They were all smiling insanely as though it was all really funny for them in their own sick way but I was not amused. Then the invasion of my mind ended as abruptly as it had begun and everything went back to normal again. I was physically shaken by it all and stayed that way for weeks and weeks afterwards until eventually I thought fuck it, if they are going to kill me then they will have to do it as there was no way I was going to stop in my mission to expose them now. I struggled on with my plight.

So with no job and no other commitments to occupy me, I spent a lot of time just sitting in the flat, watching TV and listening to the radio. They would talk to me often, commenting on my thoughts and my emotions, sometimes saying positive comments, sometimes negative. This in itself didn't stress me too much...I was used to it by now and had become embroiled in it. But it served to keep me focussed on the BBC and their conspiracy against me. At this time there was a battle going on within the BBC over me: some, (mostly the women), were for me and trying to protect me, some were against my plight. I think they must have known that I was the man for the job of successfully beating them somehow in the end. My knowledge of this helped me to realise that maybe I could take on those who were against me. But I was at a loss for how to go about it at this time. While I was deliberating, the BBC

THE DARK SIDE OF THE BBC. A DISTURBING TRUE STORY.

had already landed in force in Birmingham. The new code-word that they would say as they passed me in the street was "video" because they were supposedly making a video recording of me, whereby I was safe from any violence but everyone was allowed to take the piss out of me continuously until, as they hoped, I would eventually end up in a state of paranoia. However, I started replying "book" whenever I heard them, just to defy the video conspiracy, whereby letting them know that I would never submit to their invasion of my person and that I would always fight them off as a book and from this grew the idea of actually writing a book as a response to their videoing, in order to expose them, along with anyone else who wanted to involve themselves in the 'dark side of television' that operated against me. A book which would outline the BBC's activities and their abuse of my privacy over the years. The very book which you are now holding in your hands.

THE DARK SIDE OF THE BBC. A DISTURBING TRUE STORY.

THE BIRTH OF THE BOOK.

I went to a Government small business start-up scheme based in Ladypool Road, Sparkhill and got a grant of £40 a week to ply my trade as a writer . There were two BBC stooges there but that didn't surprise me and I wasn't going to let it stop me. They seemed to be pleased at my new approach anyway. So I set myself up and began writing my story, only by hand at this

THE DARK SIDE OF THE BBC. A DISTURBING TRUE STORY.

point as I didn't have access to a typewriter in those days. I wrote about fifty pages over three months until I began to get pissed off with constantly having to make up what I was doing on the timesheets that I had to hand in every week and started to tell the truth: that I was working 24 hours a day, 7 days a week against a BBC's invasion of privacy video against myself. When I came clean and they realised what I was doing (generating evidence of an illegal BBC controlled video), I was booted off the scheme as a matter of urgency, but the circumstances surrounding this are a little bit dodgy. They claim that I was sacked four weeks prior to the date that I really was sacked. This is because I had started sending in the timesheets that showed I was working against a BBC invasion of privacy video four weeks before I was sacked so the management back-dated the date I was sacked in order to exclude my timesheets that mentioned about the BBC. They did this in order to conceal the fact that they were trying to silence me, but they weren't going to succeed as the evidence I have shows. I sneaked back into the offices after I was 'sacked' and made photocopies of the original time sheets. Anyway, further evidence of their part in the conspiracy is provided by the fact that the woman who fired me was promoted soon after this incident. I think I was told her name was Pam. I can show anyone who wants to see, the dates on my time sheets, 23rd and 30th of August '92, show I was still on the business scheme well after I was supposed to have been sacked on 31st July '92.

Unfortunately, my enterprise as a paid writer was over and I had to forge on without finances but forge on I did.

THE DARK SIDE OF THE BBC. A DISTURBING TRUE STORY.

THE PHONEY MEDIA WAR.

It was early in 1992, just after Christmas that this so called 'war' really started to take off.

It basically was the BBC operating from London on the one side versus the Evening Mail Newspaper and Carlton TV operating from Birmingham on the other side. Then there was little ol' me in the middle who was to be the victim of harassment and abuse from both sides, as it was me that was implanted with BBC cameras and microphones, which everyone in Birmingham was encouraged to make comments into when they came into contact with me. When I was said to be with Birmingham's media, then London would harass and abuse me, but when I was said to be with London's media, then Birmingham would harass and abuse me. It was all designed to destroy me basically after I had been betrayed by our local paper the Evening Mail, who trade under their own self-description as 'The People's Champion'. Yeah! Says who? They had by now decided to turn against me the victim, the underdog, instead of taking on the might of 'the dark side of television'. I was in a no-win situation from then on and this slowly began to dawn on me over the coming months. There was still quite a lot of

THE DARK SIDE OF THE BBC. A DISTURBING TRUE STORY.

people who were opposed to this total abuse of power and they made their opinions heard both via the microphones I had been implanted with, which were then widely broadcasted telepathically and also via the ever growing 'grapevine'. I still had a large following of helpers from both London and Birmingham and this began to expand even more so now as people sympathised with my double-jeopardy situation. I went through a long period of confusion and anger at my enemies until I did the only thing left possible. I started to fight against them all.

Win or lose, I just didn't give a fuck anymore! Then one day around June 1992 , a BBC girl on the TV told me "It's an invasion of privacy".

It was true; the BBC had been breaking the law in their monitoring of me and making videos filmed through my eyes to be shown in other people's minds, usually broadcasting my thought and my vision almost continually for the most part. I strengthened my resolve. I was determined to beat the BBC, but at this point I was naive in my assumption that they were the only bad guys. I thought that other media organisations would aid me in taking a stand against them but they had all gone the other way hadn't they? They were all against me now!

I used to smoke cigarettes at this time and so the BBC would instruct people to cough openly in my presence as a means of direct provocation. Later on, this coughing was always used as an invitation to a fight and if I didn't fight the person that coughed at me, I was ridiculed and abused, so as you can imagine I started to get into quite a lot of fights. Most of which are hardly worthy of mention, usually just one punch wonders but as I had to fight my way around London in order to get some well-deserved respect, so now I had to fight my

THE DARK SIDE OF THE BBC. A DISTURBING TRUE STORY.

way around Birmingham again for the same reason. Two of the battles I got into were in 'The Nest' public house on Swanage Road, Small Heath and the opposition were both bigger than me. The first one I can't remember his name but he was well known as a bully and not particularly liked by most people who knew him. Because I was telepathic, he thought I was weak and one day, the first time I ever saw him, he just came over to me as I sat on my own and told me to come into the toilets with him. He was a big lad, but I followed him into the toilets anyway, knowing that there was going to be an almighty fight. I had an adrenalin rush as I walked in and sure enough, he started throwing punches. He was as strong as an ox but I wrestled with him furiously until we both ended up outside the toilets and about 10 yards up the corridor by the lounge. The fight was being broadcasted live to everyone in the pub at least but probably further afield and they all came out to watch the spectacle, so I just had to win or face humiliation. I grabbed him by the hair when I felt him weaken and punched him about twelve times in the face with power and at speed. He didn't return fire so when he brought the battle back down to a verbal level, I just said to him, "yeah mate sure, your safe". Then I walked out of the pub on my own because he was with his cronies and I couldn't fight them all together if they had all started. As I left, I heard one of the local lads shout out loud to the whole pub, "and Tony won!" I got a lot of well deserved respect for that little incident The BBC made this bully make another appearance two days later when I was in a friends car waiting in a petrol station on Coventry Road, Small Heath. He arrived in another car at the same time but this time he wasn't so bad assed. He was smiling this time so as not to provoke me but when I gave him the daggers look, because I knew it was a conspiracy, he turned away

THE DARK SIDE OF THE BBC. A DISTURBING TRUE STORY.

instantly from the eye contact. A few days later I told my then next door neighbour Gwen and her daughter what I had done and Gwen's daughter replied, "yes I know, and he's so big as well". I knew then that I wasn't being taken for a mug in the grapevine like I was in the telepathic conspiracy. Me and Gwen later on had a brief fling, but that fizzled out after a while and she soon became under the control of the BBC, though I don't think she was too happy about it, even though we both occasionally laughed about the situation. The other big fight I had in the nest was against Barry Ryder, another big lump whose missus was even bigger. She started it really because she was into 'the dark side of television' basically, and was out to cause trouble. She started an argument on me because she thought I was too drunk to defend myself. But I wasn't! When her boyfriend Barry joined in the argument, it turned nasty and we both ended up brawling outside. Mr. Ryder eventually backed off because I was just too fast and powerful for him but I had had enough drink in me to want to demolish him for his crimes against me. I stepped forward to give him another good thumping but I stood awkwardly on my right foot and tore the ligaments in my knee, going down onto the ground in the process. Barry himself didn't kick a man when he was down but the gaffers son at the time did. He was one of Barry's cronies and he was that brave! Luckily enough he didn't have the spunk enough to mount a serious attack and they all retired back into the pub to continue with their stopover. I got to my feet but I couldn't walk on my right leg so I hobbled to the gutter, picked up a house brick and proceeded to put it through the window. I then hopped quick time to a nearby parked car and hid behind it as I watched the gaffer's son come out and give it large to no-one in particular, before scurrying back inside before he could find

THE DARK SIDE OF THE BBC. A DISTURBING TRUE STORY.

me. He didn't want to find me on his own! I went to East Birmingham Hospital the next day where my accidental, self inflicted injury was diagnosed as torn ligaments in the knee. I passed the nest pub on my way home from the hospital on crutches and who should I bump into waiting outside there? You've guessed it, Barry Ryder!

He was smiling when we met so as not to look threatening, basically because I had given him one helluva black eye on his right side.

STEVE WRIGHT IN THE AFTERNOON, RADIO 1.

THE DARK SIDE OF THE BBC. A DISTURBING TRUE STORY.

While the media war was ensuing, I had taken to listening to Steve Wright in the Afternoon, on Radio 1. Steve had always been on my side; he loved me, and always said good things about me from the radio and into the microphones. At one point during this episode he said "We'll get him a girlfriend, but he must wear a condom". Obviously I sat up and took notice of this comment, and the next day I was out shopping with my mum in town when I saw a beautiful girl staring at me and knew that she was the one he'd had sent for me. I could see she was looking at me, but as we came close my mum made a bit of a comment, something jokingly sarcastic about her having a 'nice ass'. Of course when my intended heard this, she backed off and walked away. And it was me that got the blame, although I wasn't bothered really! It was probably all going to be media controlled anyway. Later on Steve Wright made a comment on this, although I can't remember his exact words. They were something like "She wouldn't have let you anyway", and this was definitely true after my mum had made her funny comment. Love you Mum!

Then there was Cathy Walsh. This was a girl I met at a bus stop outside the back of the Grand Hotel, as it was called then. I saw some guy pestering her and went over to get him off her back. I must have been broadcasting because he only went up to her when I arrived, and he scooted as soon as I approached and snarled at him to leave her alone. On reflection, he was probably one of the media rats that were on my case from the Evening Mail. So this girl and I started chatting, and we got on the same bus together. It turned out, much to my amazement, that she was the sister of Paddy Walsh, who I'd known for a while, though not particularly well. When this came out I

THE DARK SIDE OF THE BBC. A DISTURBING TRUE STORY.

thought I'd better not try anything on in haste for a date and we parted amicably. But the BBC had other plans, and the next day it was engineered that we bumped into each other again. I knew it was a set-up when I saw a group of BBC women watching us from the background. I saw Cathy frequently for about three weeks after that; she used to come round the flat and we'd chill out and talk for a couple of hours before she headed off. Steve Wright was getting impatient, and he made it obvious he wanted me to sleep with her. I used to have him on in the background while we were chatting and he'd make impatient comments about her, like "and she just sits there!". But after a while things fizzled out because of the Evening Mail pressuring her to turn against me and we stopped seeing each other.

I bought another car, a Ford Cortina, just a little runabout. The police didn't used to like it though and one day they pulled me over and did a full check on the car. I got done for the lights, indicator, tread on the tires, whatever they could find, and was summonsed to Solihull Magistrates Court. But this process was going to take a few months before my final appearance, and in the interim the BBC began to telepathically broadcast their 'voices' into my mind, the only one they hadn't broadcasted into so far in the whole of Birmingham. They were probably thinking that I was going to expose them at the court case, but the thought hadn't even crossed my mind at this point, they were well ahead of me. Not all of them were abusive: rather, they were intrusive. Some of them were even complementary. But the fact that I was hearing these broadcasts made me realise how the BBC had been influencing other people around me. Now I knew! I had never known how they were manipulating people and getting them to react in certain ways around me, but

THE DARK SIDE OF THE BBC. A DISTURBING TRUE STORY.

now I did know. They were using telepathy, not just on me now, but on everyone. The BBC themselves were telling me how they were doing this. They would send messages of a technical nature through the TV usually, telling me how they bounced light and sound off of radio waves, and many other methods for implementing their telepathy. I was working it all out for myself anyway, which is probably why they began educating me on the subject. Some of them even began to call me a genius, though I don't know about that. Steve Wright also confirmed that the BBC had implanted microphones in me to pick up my thoughts, and everybody else's verbal comments around me, by saying "BBC microphones" to me. While I was listening to Steve Wright I also used to have another radio station playing, BRMB, in my bedroom, one of the local Birmingham stations and as I moved between the rooms I noticed that the abuse was now largely coming from BRMB: They would make some facetious comment and then I'd walk into the living room, and Steve Wright would counteract this with his own positive comments. This situation went on for months and months, and my mind was feeling the strain a little, though not in a big way. But things changed one night when I was having a drink with Pauline, my girlfriend at the time, in my flat. I was in a characteristically chirpy mood and I raised my glass to the TV that was switched off at the time and toasted the Beeb, exclaiming "touché BEEB", something that is usually said by a sword fencer when his opponent scores a point. They must have loved that because the next day I was listening to Steve Wright, lying on my bed and he said "and he gets a well paid job!".

The meaning of this didn't sink in fully straight away, but he was saying that the BBC were going to give me a well paid job as a means of ending all the

THE DARK SIDE OF THE BBC. A DISTURBING TRUE STORY.

telepathy and the criminal conspiracy that was operating against me. So the day after, I went to the employment office to sign on and this girl there stood up and called out "Help him!". She was a BBC good guy. Of course my ears pricked up and I soon realised what was going on, normally always working on hindsight. This was the day of my hearing for my vehicle offences at Solihull Magistrates Court, and I went straight there from the employment office via a 37 bus. This journey was eerily quiet: I'd got used to hearing people say things about me constantly but now no-one said a word. They were all on my side now and willing to help me end the BBC invasion by getting a well paid job working for them. The girl on reception at the court acted a bit moodily towards me but I wasn't going to take that bait, I just ignored her and headed up to my courtroom, still smiling to myself. By the time the beaurocracy had been dealt with and the charges read out, I was feeling rather contemptuous of the whole situation because of what the police had done to me on purpose and when the magistrate asked me if I had anything to say for myself, the question really niggled me and I frowned and said "no". But as soon as I frowned, I felt my psyche drop downwards angrily as I did so and a charge of electricity was released around my head, (just like the time in the Citizens Advice Bureau). I felt intuitively that something had turned against me, and I'd lost the goodwill of the BBC. One of the bastards there must have thought "No, he's lost his temper...no job for him" and broadcasted that to all of Birmingham. As I left the courtroom I was feeling dazed and wondering what had just happened and my instincts were confirmed because outside there was a van with a bloke sitting in it who glanced up at me and then shook his head sadly, as if to say "You blew it kid". My opportunity to end all hostilities had gone.

THE DARK SIDE OF THE BBC. A DISTURBING TRUE STORY.

Knowing that I was in for some moody comments on the way home I didn't feel like sitting on the bus surrounded by people. Its a few miles back to Small Heath but I was used to walking, and it would give me the space I needed to think. When I did get back to Small Heath I bumped into a kid I knew, Mickey Shead, who said straightaway, without me telling him anything, that the only reason I got pissed off and lost my temper was that I'd had to go all the way to Solihull for the hearing. A well meant guess and half way to the real reason. So people in Birmingham knew what had happened already.

Back in my flat, I turned on Steve Wright and noticed that I was getting some real bad vibes from him. This confused me, because he'd always been on my side before. Why was he giving me grief now? I was wondering. But he answered me and said "That was then, this is now!". Obviously he was pissed off with me because it was he who tried to help me out with the well paid job, but then I supposedly blew it. Though in reality, I was sent packing by some BBC turd who didn't want the hostilities to end. I took the picture of Steve Wright I'd cut from a magazine off of my radio then...if he was going to disrespect me, I wasn't going to honour him in my funny way by having his picture stuck onto my radio!

A short time after that little incident, I met a girl through the 'heart search and romance' column of the Evening Mail Newspaper (of all places). Her name was Tracey Clarke and she was a bit younger than me but quite attractive I thought. We had a brief affair over some weeks before she turned against me due to the 'telepathy' and the cancerous conspiracy against me. She was easily turned I thought so I didn't worry too much about her when she ended our relationship until she contacted me again by letter to tell me she was expecting

THE DARK SIDE OF THE BBC. A DISTURBING TRUE STORY.

my child. We arranged to meet at her place in Lee Bank in Birmingham and when we met she told me she really was expecting my child and added "there! That will stress your mind out a bit more!" I thought that was really nice of her but because I knew she was trying to stress me, I didn't believe her and just put it down to the fact that she had turned on me in order to 'get in' with the BBC conspirators. It turned out that she did give birth to my son Jake Clarke but I lost contact with her eventually due to her treacherous behaviour. I thought about my son Jake for a long time until I vowed to see him after I had freed myself somehow from the dark side of television I was shackled to. It would take a long, long time.

     Around this time there was a police set-up involving the use of the force helicopter. I was walking up Monica road in Small Heath when I looked up into the air at the sound of a helicopter hovering above. It was a Police chopper all right and there was a white van parked underneath with three mean and serious looking chaps standing by the open back doors looking as if they were going to drag me into the van and give me a good hiding. I sauntered past without even batting an eye-lid and as I did so, the three of them having failed in their task to send me paranoid, proceeded to slam the doors shut and drove off in a huff. But when I got home and was listening to my radio (radio 1), the man on the radio said "and you went there", implying that I went para' at their earlier attempt to send me there, to which I just replied that that was bull-shit! I wasn't scared of the big, bad BBC.

     When I appeared at Solihull Magistrates Court previously (the time when I was blocked from ending all hostilities with the offer of a job with the BBC), I was fined for the driving offences I pleaded guilty to and they totalled

THE DARK SIDE OF THE BBC. A DISTURBING TRUE STORY.

some hundreds and hundreds of pounds. There was no way I could afford to pay such a penalty and so a warrant was issued for my arrest for non payment of fines. After some weeks of receiving letters from the Courts to that effect, I eventually decided to hand myself in to Police at seven o'clock in the morning so that I might get the matter dealt with that same day. To cut a long story short, I ended up in Winson Green Prison for a period of three months of which I would have to serve six weeks with good behaviour. Considering my situation, that appeared to be a nasty piece of work. I got on with it though but it was the hardest six weeks of my life. All the prisoners were telepathically controlled to unite against me as my own thoughts and emotions were laid bare to be used against me and the Prison Warders wantonly engineered situations where other prisoners would be set up in position in order to see if I would fight them. I did not feel threatened by these situations really as the other prisoners seemed to be more worried than I did, (probably because they knew what was going on, whereas I didn't) and so not a lot evolved from them. It was the constant verbal harassment I experienced that was the ignition to the fuse. You might think that would not be a real problem to someone locked up in the confines of a secure cell but this was not so in my case. As soon as I was in Prison custody, I was allocated a job in the kitchens within the confines of about ten million other prisoners. Now any prisoner would give their right arm for such a lucrative job but it was obvious that I was put there so that I did not have the near privacy or near security of all day lock-up but was in fact thrown into the largest gathering of convicts allowed in any Prison. My teeth seemed to be permanently clenched with anger all day every day for the time I spent in the Kitchens. I do not know many people that could have endured the verbal

THE DARK SIDE OF THE BBC. A DISTURBING TRUE STORY.

hassle and apparent danger that I had to go through without any means of escape as I did and so in the interim, I just had to smash somebody up. It was so very hard to focus on anyone that invaded me because they came and went in a flash from all directions and from so far a field that it was overwhelmingly confusing. It was on one occasion when the talk turned to someone actually fighting me or at least making me feel threatened with violence that I was actually confronted by some real tough guy. He appeared right in front of me as I tried to concentrate on the work that I didn't even want to do by now and looked into my eyes and said the word "me", right into my face. Then he was gone again. I was so taken aback by this that in my latent confusion at what had just happened, I became so very angry and frustrated when I finally realized the implications. The fact that he had a smug grin on his face as if he was dealing with some prick made the matter a whole lot worse and I knew that I just had to destroy him in front of everyone. I couldn't even remember his face as I walked the entirety of the kitchens looking for him; I just remembered that he had red hair. I didn't really notice the silence that enveloped the workplace at the time because of my intensive anger and confusion but they were all aware of what was now inevitably going to happen due to the fact that they were all tuned in to my brain and were experiencing my reaction at the time. The bluff was called and so it now had to be backed up. I could not find any red hair anywhere in my search and so I ventured out of the confines of the kitchen area and into the vastness of the servery area. From a distance, I could see the figure of a man with red hair and so I made my way towards him. He stood alone and apart from the others at a distance of about ten metres so it occurred to me that he had been waiting there for me as

THE DARK SIDE OF THE BBC. A DISTURBING TRUE STORY.

I advanced and was anticipating a rumble. As to whether he had been so situated by controlling voices in his head or by the wardens I cannot be sure but I suspect it was by the voices. Now he was calling my bluff it appeared but the fact of the matter is that I do not bluff. I knew I could be in danger of retaliation from other prisoners as I was the odd man out but it didn't worry me. I would go down fighting. On the other hand, I did not really think anyone would interfere on his behalf because I was the underdog and often felt respected for my convictions. I went up to him as a matter of necessity now and beckoned him to a place a bit further afield to where we would be more out of view from the warders. He knew I was about to attack him and was not happy with what he had incurred to the point where it would be fair to say he was frightened. But it was too late now, he had overstepped the mark and the onus was now on my shoulders as to the outcome. I knew the whole prison was 'watching' and in such a tough regime my only option was to destroy him now or face the humiliating consequences for the rest of my sentence. I was angered enough anyway so I just thought of him as a fool as I questioned him as to why he did what he did. I gave him the opportunity of first blood as I often did in those days because I was so confident I could take him. He tried to back out of his own creation so I knew I was just going to have to batter him now anyway while I still had the chance. He should have capitalised on his opportunity that I gave him because he was now wide open to my inevitable attack. The first punch I unleashed sent him stumbling backwards into full view of the wardens and so I quickly followed up with ferocity of lefts and rights to the head in order to finish him off within the few seconds left allocated to me. I had to teach him a lesson he would not want more of. I was then surrounded

THE DARK SIDE OF THE BBC. A DISTURBING TRUE STORY.

and restrained by the wardens who did not seem too surprised by what had just taken place. In fact, when I was allowed to return to a standing position held on to by about three wardens, one of them turned to face the audience and said aloud, "he's got a good right hook". Then I was led off to the block where I was charged with assault and told I would appear before the Governor the next day. I was kept overnight down the block and up on adjudication the following morning. The Governor was all smiles as I stood before him knowing the gravity of the situation I had overcome and seemed to know that the telepathy had stopped because of it, temporarily anyway. He asked me as to why the incident had occurred but instead of telling him forthright about the use of telepathy in his prison to which he could have taken measures to dispel, I just told him that it was over nothing really and was only over a silly argument. I didn't know enough really to feel convincing. No sooner had I said this, I saw one of the warders on duty in the room we occupied suddenly rattle vigorously for a second as though he had instantly been overcome with a vindictive controlling voice in his head urging him against me. I had not exposed the use of telepathy against me and so they were back in force and on my case again. The Governor himself let me off with the charge against me on the basis of it being my first offence whilst in Prison.

The rest of my sentence was a bit easier from then on because it was obvious to them all that I was a fighter who had no qualms about sticking up for himslf and one who knew no fear. On the last day of my sentance, it was engineered that one of the cons from the servery would try to intimidate me as I waited for my breakfast with a killer stare in order to see who would back down first. I connected to his stare straight away as it was obvious what was going on and

THE DARK SIDE OF THE BBC. A DISTURBING TRUE STORY.

returned his stare with mine. Just as I felt the need to go over to him for a confrontation as he stood with some other smaller guy who appeared to be acting as a biased witness, he averted his eyes suddenly and they both retreated to the obscurity of the kitchens behind them. It seemed I had won the final battle of conflictions. As I was in the process of being discharged from prison that same morning, one of the warders asked me if I had any complaints about my stay in custody. I just casually replied to him that I didn't. I was still alive and kicking so I wasn't even bothered about moaning about it now. He beamed with a respectful smile at my lack of concern and announced openly, "Tony's all right, and he'll be back". He was saying to them all in Prison talk that I would not be pussified and would remain bad enough to warrant a return to Prison of which I would not show any fear of. Unfortunately, he was so bloody right.

When I was returned to the streets of Birmingham, I was met with mixed feelings of fear and anger and joy at what had happened inside from the Evening Mail and the BBC. I was only doing what was expected of me but it seemed the Evening Mail guy that stood across the road with a pretty girl with him as I left the Prison was fuming with rage at what had happened inside and was making his feelings more than obvious to me. I didn't know if he wanted to fight me or not but he looked so pathetic as he stood up to his full height of about two foot nothing that I just had to smile back at him in all honesty for his own safety and so did the girl who stood next to him. She was probably Evening Mail herself but could have been BBC I suppose. All was quiet as I walked through town to catch a bus home and I finally arrived at my flat some time later where I switched on the TV to see Richard and Judy on the box. The

THE DARK SIDE OF THE BBC. A DISTURBING TRUE STORY.

first word that came out of Richard's mouth as he appeared to be watching me was the word "respect".

JANET AND BERYL CLARKE AND CLAIRE JENNINGS.

It soon transpired that the BBC were on my case again and this time things were getting serious. They had researched into my past and made contact with my first two great loves from my teenage years: Janet Clarke from Harbourne was the first girl I ever loved, and she was approached by the BBC first of all but she didn't want to get involved with what they were doing to me so they had to get her mother involved. At this time I'd taken to walking around town with my Sony Walkman on, so I couldn't hear what was being said to me by any passers-by.

Of course this meant the level of abuse would be much worse, because they knew I couldn't hear them, but I was oblivious to that at that time. But one time as I was turning the tape over, I looked up and saw Beryl Clarke, Janet Clark's mum, standing with a BBC woman on New street who had brought her along to wait for me. As I approached them, the music on my Walkman ended

THE DARK SIDE OF THE BBC. A DISTURBING TRUE STORY.

and I heard Beryl say the word "No", because they had asked her if I still wanted to fight. I wasn't going to get involved in the situation so I just walked past without talking to her. On reflection, Beryl should have said "yes" because I detected the level of cowardly abuse began to escalate to such an extent that I had to call out "yes" myself in order to control the situation again. From then on, Whenever I heard the word "no" being said, I would quickly have to reply "yes" , that I do want to fight, just so that I could keep the pathetic mob at bay by keeping them all afraid of me. Soon enough after that, all the good guys would also say "yes" because they knew I was capable and it kept the 'cowardly rats' at bay.

So that was their attempt on my first love. Soon after that, they brought in my second great love, Claire Jennings from Druids Heath. I was sitting in Billy's Bar in town with some of my old workmates from Initial UK when Claire walked by the window accompanied by some BBC fella. I noticed her, but I didn't rush out to speak to her, I just sat there looking and waiting as I knew it was a set-up and in a minute or two she came inside and sat down with this BBC bloke. I eventually went over and said hello to her.

We talked, and I could see this BBC guy was watching me in a nasty way but Claire, still smiling at me, not-so-subtly pushed his face away to the opposite direction from me, knowing that it would be a bad idea for him to piss me off. The old volatility's of our relationship soon resurfaced though and I lost my temper with her because of her involvement with the BBC. They left soon enough and I went back to my mates. I saw her three other times in quick succession soon after that, and it became ever so clear that she'd been got to by the BBC. The fist time we bumped into each other again was just down the

THE DARK SIDE OF THE BBC. A DISTURBING TRUE STORY.

road from Billy's Bar again. I was with a coloured friend of mine, Glenn Johnson from Northfield and she was with this same middle-aged BBC fella. We had a brief conversation before she declared into the microphones "I'm with him", in a futile attempt to make me feel jealous. The second time I saw her in New Street with another, younger BBC stooge, from the opposite side of the road where she stumbled in panic when she 'visualised' in her head that I was watching her. The final time I saw her with her young son in the Pallasades. She told the kid to wait outside while she went into a shop, obviously wanting me to follow her in and talk to her on her own. But I knew by then that she'd been got to so I deafed her out completely. Then I saw three BBC women standing in a circle before me with a 'resigned to defeat' smile on their faces because they knew then that I was not going to run after Claire and make a fool out of myself like they were hoping I would but I just wasn't interested by now. Unlucky babes! But I still love ya!

THE DARK SIDE OF THE BBC. A DISTURBING TRUE STORY.

TAKE 'EM TO THE CRIMINAL COURTS!

So what was my next move? This was the thought on my mind now. I was at a loss for a plan of action until a passer-by in the street told me "It's gotta end in court". I knew this was the right idea almost straight away and the sign on the Coventry road Roundabout proclaiming "The Advocates" only served to solidify this certainty in me, though seeing that sign then was just a coincidence. I began to dedicate all of my waking hours to formulating a plan to achieve this but went through a long period of uncertainty as to achieve my goal in the criminal courts or the civil courts. I had already tried and been rejected by the BBC-controlled Citizen's Advice Beaurea to take them to the Civil Courts. The only other avenue was to get myself arrested for some deviation from the straight and narrow and use the trial to expose the illegal activities of the BBC in the criminal courts. This was the only way I could force the legal system to

THE DARK SIDE OF THE BBC. A DISTURBING TRUE STORY.

take action against the BBC as they had so far been encouraged to turn a blind eye. This was towards the end of 1992 and I was walking up Hob Moor road deliberating to myself in my mind, criminal or civil, when one of my friendly neighbours exclaimed in a rather alarmed voice "criminal".

She was telling me to take them to the criminal courts, which is what everyone must have known, except me. It took a few weeks for it to sink in and for me to really believe it was the right way forward but when it did, there was no stopping me.

I headed into town one day with Dessie Mulcahy and my trusty camera in tow. I had decided that it would be a good idea if Dez took pictures of my arrest, just for the record. I walked out of Waterstones with two books under my arm, but they weren't taking the bait and didn't even come after me, so I just ended up giving the books away to some girl I passed in the street who seemed amused at the situation, so I knew it had all gone telepathic again. Rackhams was next on the shopping list and I made off with three fur coats tagged at £150 each. Again, no joy because the alarms wouldn't go off as I walked out of the exit doors for some reason. Dez was waiting outside.

Conspiracy or what? They must have known I was coming and turned the alarms off. This wasn't going as planned so we digressed to the Temple Bar for some lubrication and reconsideration. I needed alarmed goods to get myself arrested, that was for sure. We deposited the animal skins in the bar – the barman must have made a pretty penny off them that day. The clothes store we tried next was more successful.

I walked out of a clothes store on New street with some alarmed goods that sent the bells-a-ringing as Dez took pictures of my capture. I was detained by

THE DARK SIDE OF THE BBC. A DISTURBING TRUE STORY.

the manager and the police called. Only one Police Officer came in attendance and when I mentioned to him my exposure of the BBC, he said to me "surely that's a matter for the Civil Courts isn't it? " I then explained to him that I had already tried that route but the BBC had blocked me, so as it was a criminal matter, I was taking them through the Criminal Courts. The staff were quite keen to have me arrested, but the police thought it would be more pertinent to take me into the back room and kick my head in. In the end neither of these happened and no charges were pressed, the attending Police Officer made sure of that as the police were covering up for the BBC for some time now and not just by turning a blind eye as I have mentioned earlier. I stumbled home and repaired to my bed where I slept on the problem. I awoke with an urge to visit the local DIY store in Small Heath in order to try again to get myself arrested for shoplifting. The telepathy against me in my head was strong that day, but I was so determined to complete my mission as I walked towards Texas Homecare that I lost my temper with the BEEB again. When I did, a woman's voice in my head said "do it that way!". I had to appear to be mean enough and fearless enough to take on the big bad BEEB! From then on it was widely broadcasted that some of the BBC were helping me to succeed. I went into Texas Homecare on the main Coventry road to get myself arrested for shoplifting and several voices urged me in my course of action but I was content to take my time and find the right item. I browsed the store nonchalantly before my fancy took to an alarmed pair of curtains. I've no real reason why I chose them!

THE DARK SIDE OF THE BBC. A DISTURBING TRUE STORY.

The alarms screamed as I first ambled my escape through the automatic doors and then I stopped completely when I got outside in anticipation of the inevitable moment of apprehension.

The security guard demonstrated an uncommonly contemplative approach to his job and when he'd arrived and I'd told him it was "probably me you are looking for", I was brought back into the store. The police made their way dutifully to our liaison and escorted me back to Acocks Green Police Station, where I and my belongings were booked in. When I mentioned my exposure of the BBC, the desk sergeant said "and you're also writing a book aren't you?". I replied that I was and he wished me success in my venture by declaring,, "So the word is 'successful' then" . Not all coppers are bastards! Mind you, they were probably only being nice to me because it had been widely broadcasted that the BBC were helping me to succeed.

I chose Ian Gold's solicitors firm, who were based in Moseley, to represent me and found myself in the capable hands of Caroline Salvatorie. I knew from a past episode (when I was arrested for all the motoring charges and dealt with at Solihull Magistrates Court where I lost the job offer from the BBC) that Ian Gold himself had declared that he would take on my case in Court to help me expose the illegal activities of the BBC. The case was to take place in Solihull Magistrates Court again and I was confident things were heading in the right direction. The doctor who pre-interviewed me to make sure I was fit to be questioned advised that it was probably a good idea for me to "stay away from television" in the interim. I said in my tape recorded statement that I only committed the crime in order to expose the illegal activities of the BBC, but I was given the wind up by the interviewing WPC who tried to convince me that I

THE DARK SIDE OF THE BBC. A DISTURBING TRUE STORY.

had been set-up by the BBC. I lost my temper at her piss-taking and concluded the interview saying that I didn't want to say anymore about it.

Soon enough, I was placed in a cell and left to contemplate the future as my thoughts were being broadcasted to everyone for miles around continuously. This is when I progressed from sending and receiving audio telepathy to sending and receiving audio *and* visual telepathy.

This was the first time I ever knowingly received visual broadcasts into my head and believe me it was meant to terrorise the holy Jesus out of me. It came in the form of a gigantic snakelike, barracuda-type creature that relentlessly and savagely attacked the defences of my mind. It wasn't just that I could see this monstrosity attacking me, but I could actually feel it as well.

I could feel its ferocity, its demonic hatred, and its violent desire to destroy me as it continuously lunged at me again and again and again, broadcasted to me inside my mind. It was scary, It could have been *very* scary if I hadn't have done what any true fighter would have done!

I immediately burst into laughter like a madman and jumped to my feet, using the power of my mind to fight back against this onslaught by focusing all of my mental strength against the creature. I was in a state of controlled madness! I even threw a few punches at the fresh air, shadow boxing at it, until I reached a point where my confidence felt so good that I began to taunt it by chanting to myself, " barracuda, barra- barracuda, come on, come on! It did come on too, on and on at me as I continued to focus on it, taunting it, laughing hysterically at its violent attempt to frighten me to death. Let's face it; I went a bit fucking mad! But that's why I won. Won, won . The attack lasted approximately four minutes in all, then it just stopped completely in an instant and I simply

THE DARK SIDE OF THE BBC. A DISTURBING TRUE STORY.

returned to normality. They obviously realised that there was no way they could beat me from then on I reckon. Either that or they just ran out of film or something! (laughs). Anyway, when that ended, I could hear the Police Officers outside my cell at the desk were taunting me for my thoughts, so I knew I was thought-broadcasting in a big way. The Police could have charged me and let me go by now, but instead they thought that they would amuse themselves at my expense instead. I thought "fuck this; I'm not having that all day!" So with bright shiny eyes and a wicked grin on my face, I advanced on the cell door and put my head to the open hatch to see if I could locate the position of the intrusive Police 'voices' from outside. I couldn't. So using my ears to listen and my minds eye to locate them, I began to use telepathy to my advantage by unnerving them in the hope that they would let me go. And it worked! By Jingo, it bloody well worked!

I didn't use my mouth to speak at any time at all during this little phase; instead I used my mind because that was what they were tuned into and preying on. In not so many 'words', I told them that I was not the only one that was being observed and scrutinised, but that the BBC were also watching them as well. I continued to taunt them telepathically until one of them broke down and couldn't handle it any more. I sent him paranoid! Hee Hee!

Within minutes I was unlocked, taken to the desk, charged and released. I was back on the streets again, but not in the best of moods now. I travelled on foot back to Small Heath but after only about two hundred yards, I came upon a black guy with a white girl coming from the opposite direction. When they came in close enough, the black guy, half looking at me and half looking at his accomplice declared "it's not funny y'know!" He was quite right, it wasn't funny

THE DARK SIDE OF THE BBC. A DISTURBING TRUE STORY.

but I had my first court-case pending now against the BBC so I didn't give a damn, basically.

I returned to my beautiful flat on Hob Moor road, had a beautiful cup of tea and hit my beautiful bed.

The second court case happened quite by accident really, but luckily enough it did......well!.... It happened one night when I was out drinking on the town with Dezzie's brother, Mickey Mulcahy. We were having a good time together, drinking, laughing and flirting with the women. I was driving my car that night so when I felt myself getting to feel influenced by the drink, I parked the car up off Bradford street, disconnected the battery so that I wouldn't drive home and then went into the White Swan with Mickey. We stayed there for the rest of the night in the White Swan until I got slaughtered. It was then that two Police Officers appeared and started talking to me. Apparently, one of these coppers said that he knew me and that we used to go to school together years ago. I didn't recognise him at all so I thought he was bullshitting me and just trying to gain my confidence in order to stitch me up somehow. I turned a bit nasty on him then, taking the piss out of him, as you do, in front of everyone. Eventually, we left the pub as I was still very, very pissed. A voice in my head told me to get the car quickly and drive home. I refused to listen to it at first but when it implied that the Police were going to get me, I believed the little bastard and re-connected the battery somehow and drove off home, still completely pissed, but capable. I dropped Mickey off at his parents house where he lived at that time on Aubrey road in Small Heath and Mickey invited me in for a coffee so that I would sober up a bit. Unfortunately,

THE DARK SIDE OF THE BBC. A DISTURBING TRUE STORY.

or fortunately, which ever way you look at it, it didn't work and I drove home well and truly over the drink-drive limit.

I was so pissed that I was ranting and raving out loud to myself in defiance of the criticising voices in my head. They led me to believe that there would be some kind of violence waiting for me when I got home and stone me, they were right .Apparently, this was supposed to be a 'present' for me. It was pretty blurry around me, but when I saw the two BBC guys waiting for me on the driveway up to the garages outside my flat looking very, very nasty, I believed I was in for a row, so I stopped the car halfway up the drive and got out, ready for it. Of course I was in no fit state at all to take on the two of them, but I really didn't give a damn! If these two guys were so brave enough to take me on when I was that clattered, then I suppose I had to oblige didn't I? Fortunately for them, The Police arrived within seconds to save their asses. (Laughs again). I saw them get out of their patrol car with the same shiny eyes and big grin on their faces and advance right up to me telling me that I had been drink-driving. It is so obvious now that they were following me all the time, but all I could do was to roar back at them "well, I'm not driving now am I?" They were still smiling as they put me into their car and took me to Stechford Police Station, where I promptly fell fast asleep until morning. I wasn't given bail. I was taken to Birmingham Magistrates Court at about 6.30 am that morning and given breakfast in the lockups under the Court House, of marmalade sandwiches, which weren't too bad I suppose. When we were transferred to the court locating cells, I could tell I was thought broadcasting, by the way people were reacting around me but I really didn't care. I never once felt threatened, just annoyed really. When it was my turn to appear before

THE DARK SIDE OF THE BBC. A DISTURBING TRUE STORY.

the Magistrate, I landed in the dock looking cool, calm and collected. Before the proceedings even got started, the Magistrate who was presiding on the bench waved his finger angrily at the prosecution council and demanded to know "why wasn't this man given bail?" He was siding with me now because it was common knowledge that the BBC were helping me to succeed. To this, the prosecutor dropped his head and replied humbly "too drunk your worship". That was his contribution. So after a few legal discussions took place, I was given bail and ordered to return to the Court house in one month. I now had two court cases pending and things were definitely looking up.

TRIALS AND TRIBULATIONS.

I was focussing on the theft of the curtains case being heard in Solihull. The weeks passed by into months with me having to attend Court each month to make an appearance. I was given a car to drive around in by a mate of mine, John Ellis, the same one that helped me install some furniture into my flat shortly after I first moved in. But some of the BBC still didn't like me to do well, so they got someone to steal the battery out of the car which I kept parked on the grass outside, so that I couldn't drive it. I was outraged about that, so much so that I pushed it from off the driveway to my flat and parked it

THE DARK SIDE OF THE BBC. A DISTURBING TRUE STORY.

lengthways across the middle of the road in order to block off the traffic, as some sort of protest. I left it there and returned to my flat in order to calm down a bit. I even took pictures of my one man protest and still have them today. The Police arrived soon enough, so I made an official complaint to the Police about the theft of my battery. They left it for a while before they acted on my complaint, but when someone from Stechford CID did arrive at my flat to deal with my complaint, I told him that it could have been anyone from the BBC or from Carlton TV as I believed they were both in cohorts against me. This was at a time when the BBC were trying to trick me into believing that Carlton TV also had this broadcasting capability and were also using it against me. Shifting the blame, so to speak.

The cop jumped at the chance of not taking any action against them by declaring that if I didn't know which of them it was then there was no way that the Police could take action against either of them. I had to wait to expose them at the Court Case. The whole of Birmingham waited with baited breath.

After about six months of waiting, the day of my trial arrived, the day of reckoning. It was around Christmas time, 1992. My solicitor, Caroline Salvatorie had a statement that she had written out herself, from her own personal knowledge – I definitely had no say in it - about the illegal activities of the BBC. I still have all documented evidence and the statement she wrote by her own personal knowledge of the facts is reproduced and kept on file.

I entered the courtroom early to find the judge and my solicitor, Caroline Salvatorie, deep in conversation with who I now know was a male BBC

THE DARK SIDE OF THE BBC. A DISTURBING TRUE STORY.

operative. I had just enough time to see this before being ordered outside by the Court Usher, who obviously didn't want me to hear anything of the conspiracy that was taking place inside the courtroom. I waited outside in contemplation and when Caroline Salvatorie came out and approached me again, she came wearing an extremely negative attitude topped off with an opinion that she needed a doctor's opinion on my mental health. I was so confused of what was now going on now that I allowed it to continue. So the trial was postponed and I was decanted to Doctor White in Edgbaston, who took the piss for the duration of the interview, watching me with the usual grin and eyes shining. I left in a less than happy mood as it was now slowly sinking in that a cover-up was taking place and the group of student-type kids outside did nothing to ease my troubled state of mind. In fact, their sarcy comments made me even moodier and I decided to walk back home to Small Heath in order to get my head together. I was still substantially pissed off as I reached my own backyard, when this guy walked out of the Black Horse pub on green lane, Small Heath, coughing right in front of me, the now usual provocation to a fight. I'd had enough of this shit so I whacked him one in the mouth, cutting up my knuckle and removing one of his teeth in the process. He swelled up with rage for a split second, looking as though he would retaliate against me, before deflating back into the cowardly little rat that he was and scurried off into the shadows of the early evening gloom. He wasn't in my league really now as I continued to grow bigger, stronger and faster with every passing week. I had to. My knuckle bled all the way to Michael's Supermarket on Green Lane, where I bought a dressing to stem the flow of blood, and as I walked out I heard an Asian girl cry out "It's all about money". This was in order to remind everyone

THE DARK SIDE OF THE BBC. A DISTURBING TRUE STORY.

that it's not about fighting but about money that I should be suing for. She obviously said this into the microphones to help me because the BBC must have been telepathically inciting everyone to violence against me for knocking that guy's tooth out. All the Asian girls were on my side now because they liked me, probably because word had gone around the 'grapevine' that I had dated an Asian girl for over a year and that I was a bit of all right as well. I wasn't bothered either way though as I could easily have knocked a few more teeth out!

So with the aid of Doctor White's report, the case was ceremonially dropped. The only incident worthy of mention was that the female magistrate dealing with the case called me a "good for nothing" after she dropped the case. Good for nothing! I, the man striving to free our good nation from the shackles of the BBC's mind-control! These people might have their diplomas and accredited qualifications but sometimes they really are shit for brains, honestly! Anyway, I found out years later that Caroline Salvatorie, after my interview with Doctor White, had written a letter to my own family practitioner Dr. Harrison about the 'voices', saying effectively that they were all in my imagination. The whole thing had been speedily covered up.

Both the Evening Mail and the Stetchford CID bloke who stitched me up for my car battery complaint were lurking outside the courts. The former just gave me a dirty look, but the latter made it clear with his knowing grin that things were messed up bad, though he wasn't the one involved in this cover up.

So my hopes were pinned on the second court case for drink-driving which was being handled by Ian Gold himself in Birmingham, but things were taking a turn for the worse - in fact they were going totally haywire.

THE DARK SIDE OF THE BBC. A DISTURBING TRUE STORY.

Robert Maxwell, an ex Small Heath so-called hard man, now one of the DJs at my brother-in-law's club, The Emerald, had been brought in to take over the assault on me and was attempting to enforce a nervous breakdown on me prior to the Court Case. He was operating from either the studios of the BBC or Carlton TV in either Birmingham or London, or so I believed at this stage. I was told by two women passers by from the Evening Mail that it was from Pebble Mill, the BBC studios in Birmingham, but I couldn't trust anyone at this stage though, so I didn't. This was around Christmas 1992 and the case was to be heard in March '93, so his electrical attack was intensive and relentless. He worked with a team of local lowlifes from Birmingham, but I didn't know who they were for sure, though obviously Maxwell does. He continually used the full power of 'The Dark Side Of Television' against me, using any kind of noises around me to channel them into human 'voices' that were abusive towards me, in an attempt to break me down into a frightened, paranoid wreck. He used the sound of the wind at my bedroom window to abuse me at night, he used the sound of the wind from passing cars when I went out of the flat , the sound of my own urine as it splashed in the bowl, he used everything. The attack was relentless and I even experienced 'films' being broadcasted into my head as I drifted into near sleep in my bed every single night of the week. Mostly the 'films' were of a horrific nature or of a sexually perverse nature that tried to terrorise and corrupt my mind. During this period however, the BBC seemed to have changed their position again and said mostly good things about me. Probably they felt sorry for me: because now it was the local media, who had taken over the conspiracy and were victimising me, using people on the streets and of course the now great Robert Maxwell who had been recruited into

THE DARK SIDE OF THE BBC. A DISTURBING TRUE STORY.

television. The aluminium desk-lamp on top of my TV blew it's bulb about once a week because of the electrons that were transmitting through the TV set. My television was absolutely covered with electricity. This transmission was so powerful that once when I was sitting in my living room, I could actually see the cloud of electricity surrounding the TV set. It looked dense and misty. I felt myself covered in electricity, as if there were a huge weight pressing in on my body, which was a manifestation of the evil Maxwell was emitting toward me. I could visualise him in my mind occasionally, almost like an apparition and it was this night that the voices in my head began big-time. From then on every night I heard voices criticising and abusing me, mostly at night while I was trying to get to sleep. They continued until 3 or 4 am then would ease off for a couple of hours and begin again. They were working round the clock against me, torturing me with sleep deprivation. During the couple of hours remission I managed to sleep but my mind was still tortured with vulgar 'films' that were transmitted into my head, often of being attacked by black people, some of whom I was acquainted with in Small Heath. They called these films 'dreams'. One time, I 'dreamt' of them pissing on my head, and could actually feel the hot liquid splashing down on my forehead. Maxwell was trying to make a racial issue of it, and "black" became his code word: Whenever I was called on to fight him, I would have to fight through black people in Birmingham first before I could get to him. However, there was never any real aggression from the black people in Birmingham. Mostly they felt solidarity with my situation, though obviously there were some shit-heads. There was a period of about five weeks before the court case when things were intolerable: As well as the voices, I often experienced that I was being physically hit repeatedly over the head with

THE DARK SIDE OF THE BBC. A DISTURBING TRUE STORY.

a hammer during the night. All done by electricity. The BBC still stuck up for me - once one of their guys reassured me by saying "it's bad what they do". The majority of them were on my side by now, impressed by my willingness to fight the local media and still continue to write my book against all the odds. It was at this point that Maxwell showed himself for the dirty, perverted coward he is: He simulated the act of sticking something up my butt from the safety of his 'studio', or wherever he operated from while I was standing in my front room watching the television (it made no difference whether the TV was on or off). I knew what he had done straight away, and this was later confirmed when a girl's voice on the TV said "He was raped". From then on, the word "finger" came into regular use so you can draw your own assumptions. The TV also hinted to me "it's those little Devils at Central" (Carlton's old name before they changed). All through this, they were still transmitting electrons through the TV at me. I could see them travelling through the room and into my head via my ears, sound effects included. When I went out into the street, my vision was distorted by savage and relentless blows to the eyes. As an alarmed woman passer-by said, "They're trying to cabbage him". Many people hinted to me that it was the work of the Evening Mail people. I'd stopped replacing the bulb in my aluminium lamp on top of the television because it kept on blowing due to all the electricity surrounding it and was proving very expensive to buy a new one every time. But once I forgot that I had already taken out the blown bulb without replacing it and tried to switch it on: Obviously it didn't light up, but just then a tremendous surge of electricity was sucked from the TV, sucked from me, through the lamp and out into the atmosphere via the window next to the TV. I had unknowingly blown the electrical attack away! Claire Short, the

THE DARK SIDE OF THE BBC. A DISTURBING TRUE STORY.

Birmingham Labour MP was on the TV at the time of this incredible feat and she declared "And he gets up and leaves". She was telling me to get out of there quick before the instigators of the electrical onslaught came round to my flat, but I wasn't going anywhere. In fact, I felt total relief from the electrical attack now for some 48 hours. But I knew there was no escaping them, there was nowhere I could run. I was staying right here, in my home, and if they wanted to come and finish me off, they would have to do it on my soil. Two days later the news reader Carol Barnes said "And none of them came round". Then I realised that the electrical 'line' to my brain had been reconnected, but prior to that they had hidden, Robert Maxwell and his low life cronies, they didn't have the courage to come and face me which is what Carol Barnes was emphasizing with her comment. I realised then that the surge of energy I released when I turned on the lamp had drained them completely of their electrical power over me, wiped them out. I had stumbled upon the one and only weapon that could vanquish all of their heavy electrical attacks via the Television. A £5 table lamp from Woolworth's. Now I knew they could never break me! This was fate surely.

I then remembered a dream I had when I was a youngster of about ten or eleven years old. It was one dream that I had on two different occasions, both exactly the same dream. In it I remember the feeling of power I experienced as I invaded the privacy of a man who slept on his bed, blissfully unaware of my presence as I watched him through some kind of powerful lens in his bedroom. The feeling of power persisted for a few seconds as I watched him until suddenly the feeling of power turned in to that of dread and fear. Then the dream, or premonition, ended as abruptly as it had begun. I realise now that I

THE DARK SIDE OF THE BBC. A DISTURBING TRUE STORY.

experienced the dread and fear in that dream because it was me, myself, that was in the dream all along, and looking at me now, it was a real-life premonition of things to come. Holy Moses! All this is my destiny, surely! Scary huh!

I would do anything now to succeed in my mission to expose and so put an end to telepathy, but as I said, two days later, the BBC had reconnected the electrical 'line' to my brain and Robert Maxwell returned to his sleazy seat of power, now said to be "hiding behind the BBC". He could easily have been with Carlton TV, who were themselves hiding behind the presence of the BBC in Birmingham I wrongly deduced. Any which way, he was rewarded for his long stay stint of evil with a slot on ITV's blind date series with Cillla Black, where he made a bit of a Pratt of himself as a show-off in the early 90's. It was then that I began to gather clues as to who was behind this new wave of telepathic violence: On the street, on the TV and radio, in my head, everywhere I turned I heard the name Robert, or Max, or Blondie (this Robert Maxwell had blonde hair). I couldn't believe it at first, that Robert Maxwell was orchestrating this against me. It took a long, long time to sink in because I could only work on hindsight. Maxwell and his team from the Evening Mail began calling me a 'wanker', as opposed to the good guys who called me a 'geezer' saying that I was a bit tasty. As I walked through the streets some days, the wind was channelled into voices in the hearing canal of my ear, saying "wanker, wanker'. All types of sounds were channelled into human voices in the hearing canal of my ear and from the source. When I reached the end of the road a BBC girl turned around and said "Okay calm down, calm down", telling Maxwell to ease off. But the lower classes from the Evening Mail were in control of the

THE DARK SIDE OF THE BBC. A DISTURBING TRUE STORY.

conspiracy now, operating alongside the BBC and were adamant that they would turn the whole of Birmingham against me to stop me from suing them and the BBC. The BBC had slowly returned to London now after the court case had been dropped in Solihull, so at least I was on the right track and making progress. Some of them were helping me with my mission as they often told me I was "Big", as in big-time. The Evening Mail were determined to dominate me and keep me submissive that way for the rest of my natural life. That was not in my destiny, I hoped, so I would have to stop them! It was easy enough to recognise the remaining BBC people on the streets because they still wore their characteristic smiles and shining eyes, though they were now largely on my side. The Evening Mail people had no smiles, and their faces showed only malevolence. This was the difference between them and the BBC.

I was in the Cauliflower Ear Boxing Gym at Deritend one evening, doing my training. I was feeling aggressive because of what was happening to me and I jumped into the ring, thinking of sparring by myself. Robert Maxwell soon appeared then, smiling like an idiot instead of frowning as you do before a confrontation and this half-cast guy there gave him a real dirty look, as if to say "What the fuck are you doing, you tosser?". Maxwell was just standing there, doing nothing and looking like a nobhead...he was supposed to be the 'one' in Birmingham! After two minutes he just shot off out of there without doing anything but my suspicions were getting aroused now.

As the incidents accumulated and I began to believe it was Robert Maxwell attacking me, I started to hear voices saying "Heads", i.e. Who was going to win on a real head to head confrontation, him or me. Maxwell himself was still elusive and obviously didn't want to get involved in any real confrontation. I

THE DARK SIDE OF THE BBC. A DISTURBING TRUE STORY.

saw him once on the Coventry Road in Tyseley and demanded to know whether he was involved: Admittedly my approach was rather aggressive but he was adamant of his innocence and I left it at that, after advising him that he'd "Better fucking not be". However on the way to my mum's in Yardley straight after that incident, some girls were giving me seriously nasty looks and I wondered much later if I'd blundered. I was supposed to fight Maxwell then, and I'd just let him walk away. How was I supposed to know what everybody else knew? I was always the last to know things at this stage and had to learn what was going on around me from the 'vibes' I got from people on the streets. In my flat, the TV confirmed this. "Fair play to him and his team" was said, referring to Maxwell. Because it was me that didn't start a fight, I was being dubbed a 'broken man'. I wasn't going to accept this. I'd never lost a fight as yet, not even by default. One day I got so vexed that I started to believe it was indeed him so I picked up my hammer and was about to go out and find him. Even the girl in my head was saying "Hit hard, hit hard". But I backed out. I became unsure again. This did seem a bit bizarre suddenly. Could it all be bollocks? I put the hammer down and went out to go for a walk and get some fresh air. I passed by Maxwell's Mothers house (she only lives some hundred yards down the road) and his mum was out the front, smiling nervously at me as if to say "Oh shit! Tony is going to find out that it is my son responsible for the attacks against him". Then Maxwell himself came out of the house, wearing a very nervous smile on his face. He talked to me amicably - although I knew it to be a false camaraderie. He even invited me to his wedding. Why do this when we hardly really knew each other? Why be so friendly at all? It was obvious he was trying to hide his involvement in the conspiracy. The cherry on

THE DARK SIDE OF THE BBC. A DISTURBING TRUE STORY.

the cake was when he suggested I go to Amsterdam, which I later realised he was insinuating that I needed to start sleeping with prostitutes because he was going to prevent me from getting a girlfriend. It was a well known fact that I was attracted to the women. I suspected what he was insinuating a few hours after we parted company, but obviously I wasn't sure enough at this stage to warrant a physical attack on him. It was all trickery and voices.

As March approached, the onslaught continued. Maxwell was broadcasting into peoples' heads, telling them what to say around me, telling them to cough and invite me to fight with them. But I survived. I was near the end of my tether, but I knew I had to make it to the court case. March had arrived and with it my hearing. I caught the bus into town and then walked to the courts. I heard Maxwell in my head say "Let me out of here", and knew this to be because he was frightened off now that the court case was now about to happen.

I was unhappy to see that Caroline Salvatorie, who had betrayed me at my last hearing, was there with Ian Gold's male clerk who was called Chris. They advised me to plead guilty and then mitigating circumstances could be claimed concerning the BBC and their counterparts. I didn't know whether to go along with this. I suspected that I had to plead not guilty so that I could turn the case around on Robert Maxwell and the BBC conspiracy. But before it came to my plea I heard a voice from the crowd of people that came to witness my trial say "plead not guilty" and my gut reaction was of contrariness, "Don't tell me what to say" I thought back, and promptly assured the courtroom of my guilt by pleading guilty and waited to hear the mitigating evidence. The magistrate, who had previously been slumped over his bench with an evil look

THE DARK SIDE OF THE BBC. A DISTURBING TRUE STORY.

on his face then relaxed completely and let out a pleased smile. I began to realise then that I'd blown it. Another cover-up would begin but not before the Magistrate leaned over his bench and asked Ian Gold, "why did they disallow it?" He was asking Ian Gold why had the BBC disallowed my last court case in Solihull, when it was they, themselves that had allowed it in the first place. Ian Gold himself, still trying to help me replied "not serious enough". The court adjourned and I would be sentenced later. I now had the words "not serious enough" implanted in my brain! All over town people's eyes were shining and the message was simple: You've no chance now. My only option was to make a pub crawl home. As you may expect, I bumped into Maxwell in one of them, the George and Dragon in Small Heath, (where he was originally recruited from by Television). I said to him "Okay, if it's you then let's go outside and have the row!". He just replied all nicey smiley " would I do that to you Tony?" and skulked off into the back room. The next day when I was in my Moms house talking to my brother Mark's girlfriend Ruth, I told her that I had foolishly pleaded guilty at court the previous day and she exclaimed rather urgently "are you going to do something more serious now Tony?". This sent me into a bit of a shocked state as I then realised that the conversation between the Magistrate and my solicitor Ian Gold the previous day had been broadcasted throughout Birmingham.

I then began to think about doing a more serious crime in order to implement the expose.

I went on a crime spree then. The good voices, the ones on my side, were telling me to "Rob", as in steal, and when I did, no one stopped me even though I was thought and sight broadcasting all the time. If I didn't do as they

THE DARK SIDE OF THE BBC. A DISTURBING TRUE STORY.

said, they came back with "Coward" or "wanker". "Wanker" was still the Evening Mail's buzzword but now some of the BBC picked up on it, except they just said "Hands", a disguised way of implying that I was a wanker. I was determined to prove them all wrong! Anyway, soon enough it was time for me to go back to court again for sentencing at Birmingham Magistrate's Court where I had been foolish enough to plead guilty at my last appearance. I went with my Dad on this occasion but we were not too confident of a public enquiry being ordered. It was the same Magistrate as before but this time the whole court room was like a scene out of a film. They were all acting up for the cameras, which were my eyes. My eye-sight was being broadcasted telepathically to every-one around me (I don't know the extent of the range) and they in turn would begin to act out their little roles whenever they were 'on camera'. This was all taking place within our society's judicial system. There was gutter press talk with sexual innuendos about me, but this was always counteracted by saying the word "rubbish", or "bullshit". The same Magistrate soon piped up with the word "rubbish", so at least he wasn't sinking down to Robert Maxwell's level, but he sure enough didn't order an official investigation either. He soon piped up with the words "I don't believe you!", meaning that he wasn't having any of the truth about the BBC and their counterparts although he knew for a fact that it was all true. Everybody knew it was true! He wouldn't have been able to say that if I had pleaded not guilty and went on trial. As a man once said, "there can be no whitewash at the whitehouse". Anyway, as soon as that was said my Dad jumped up and declared "that's it then", made his excuses and left me to my fate at the hands of the so called law. Caroline Salvatorie, now acting again for me wanted me to be swept under the carpet

THE DARK SIDE OF THE BBC. A DISTURBING TRUE STORY.

and into a mental hospital, so two doctors were sent to see me in order to do just that. They tricked me into believing that they could help me, when they knew for a fact that they couldn't and so I agreed to go to a psychiatric hospital voluntarily. I'd signed my own death warrant.

This information was relayed back to the Magistrate who then, in collaboration with my former solicitor Caroline Salvatorie, sectioned me under section 37 of the mental health act and I was led off in handcuffs to Hollymoor mental hospital in Northfield. The second court case had failed miserably but I was learning fast all the time mainly by (forgive the pun) trial and error.

I spent three months at Hollymoor Hospital but I continued to remain positive by thinking ahead and by going out running around the Hospital grounds nearly every day. I met a lad in there who was in the same position as me. He was also a victim of BBC telepathy and he exchanged information with me that married up with the same experiences I was going through. He wasn't as big as me though and potentially not half as aggressive as I could be, so he was well under their control. He was telling me one day what the BBC do against him and as he spoke, the radio was playing in the background (radio 1) and I heard the female dj interject with the words "oh stop moaning". This made me smile slightly at the experience but I felt really sorry for him as I knew that he couldn't fight his way out of it as I could so I told him that I would expose the existence and misuse of telepathy one day and we would both be free. His name was Richard and he was moved to a Hospital in Solihull towards the end of 1993. This in itself made me wonder just how many others were sentenced to a lifetime of misery at the hands of the BBC.

THE DARK SIDE OF THE BBC. A DISTURBING TRUE STORY.

Some while after that I even showed my contempt for the situation I was in by absconding from hospital on one occasion, to which a lad I passed on the street on my way home declared "that's the way it should be!" But thanks to my older sister Helen who helped the police when they came looking for me, I was taken back by Police later on the same day. Thanks Helen.

The drugs I was forced against my will to take were increased because of that, which was bad enough but even the doctors and nurses there were happily making derogative comments about me into my microphones which was making the situation even worse, I just wanted to kill someone at times. It was just as bad when my Mom, my nephew Shaun and his then girlfriend Hayley came to visit. It was all people around me making derogative comments about me and what I was thinking at the time, making me rise up to boiling point. My nephew Shaun saw this and said to the instigators at the BBC studios "that's enough now", but still they persisted because they wanted to push me over breaking point, so Shaun repeated again, smiling wildly, "no, that's enough now". This showed me that some people didn't find it funny to be told what to say by nasty little voices in their heads. But things subsided a bit after that, only to return again the next day, all day, every day, until a BBC male voice from the television declared "you shouldn't kick a man when he's down!" After which things quietened down a lot until I was eventually released after three months of pure bullshit. I had to pretend to the consultant psychiatrist that I was better now, that I didn't hear voices anymore etc. He knew I was bullshitting but he knew they couldn't keep me in for ever. They had had enough fun out of me at my expense. I was discharged from Hollymoor Mental Hospital on 26/7/1993.

THE DARK SIDE OF THE BBC. A DISTURBING TRUE STORY.

LYNN VALE.

It was a long time after that that I met Lynn Vale, well it was a good few months later. She was a social worker, and I chatted her up at the kebab shop by Aston University one night when I was out with Dez. She gave me her phone number, but not after saying "What's these voices I can hear?", her way of

THE DARK SIDE OF THE BBC. A DISTURBING TRUE STORY.

saying that she knew about the situation I was in. But most women liked me by now because I was becoming a bit of a cult figure, though somewhat controversial, never giving up the fight. There was a group of Asian kids standing across the road making snide comments into my microphones which were the cross I had to bear and watching me but I didn't care. They could watch all they wanted. Just don't even think about touching. So I started dating Lynn, and I was spending a lot of time at her place in Chelmsley Wood, then we would spend some time at my place in Small Heath. I was also driving her car during the day and picking her up from work in the evening. For my birthday in February '94 she bought me some new furniture for my flat: a wardrobe, a chest of drawers and a bedside cabinet. **She was quite well off and was happy to share with me as I was with her.** As I said, Lynn knew about the conspiracy and the evil ones were working on her in an attempt to use her against me. She used to hear voices telling her to say "hands" and other such words to infuriate me, but she sometimes managed to blank them out. She told me to try and do the same, but I couldn't and my anger towards the people manipulating her and me was growing exponentially. Its funny isn't it how these people who spy on others always turn their attentions to sex. Probably because they are just sleaze bags to begin with. When Lynn herself was told to say nasty words at me, and she complied, I would get very angry with her and we would sometimes end up having big rows. At this time there was an old couple living in the flat beneath me. They were called the Gameson's and were enlisted by the Evening Mail in order that they could verbally attack me from close-range. The Gamesons would let people into their flat to sit beneath where I was in mine and talk to my thoughts through the ceiling. When there were no other

THE DARK SIDE OF THE BBC. A DISTURBING TRUE STORY.

agents with them, the Gamesons would leave their radio on constantly so that Maxwell and his cronies could talk to me through it in their own twisted way, trying to make my life unbearable. Although I had confronted the Gamesons previously, nothing major had happened. This changed during a light drinking session with Lynn one night in my flat when their interference was so intolerable I took a hammer to their windows and then set to work on their door. Unfortunately I was distracted by the police asking me to put the hammer down - which I chose not to do, opting instead to tell them to piss off, and then went back to my flat. The Police were never helpful to my plight so I was losing what little respect I had for them by then. They were at the door a few minutes after me, by which time Lynn had hidden my weapon of destruction...although not particularly well. I, my hammer and a set of cuffs were soon accompanying the Police to Stechford Police Station. I was charged with criminal damage and I knew this time I had to plead not guilty and expose the BBC at a trial. I went to George Jonas & Co. this time that had represented me at one of my first court cases for motoring offences, well before I learnt that it was a court case I needed to win this never ending battle. I told George Jonas' son Stephen about the BBC, their interfering with witnesses and the perversion of justice and although he was keen at first, both he and his father George eventually plotted against me but I was soon passed onto someone else in the firm who was less susceptible to the BBC's conspiracy. I cannot remember his name but this solicitor was on my side and was willing to help me expose the whole sordid conspiracy. He even asked me "what about this book then?" And I told him that I would have a brief synopsis ready to be used as evidence in court. He said "that is good enough". I left his office in high spirits again at the thought of

THE DARK SIDE OF THE BBC. A DISTURBING TRUE STORY.

ending the whole sorry affair once again. In the meantime, I put an advertisement in the local jobcentre for a writer to help me with my book and gave Lynn's telephone number as a contact point. For a few days there was no response, so I got to feeling the BBC had a hand in that. Then one day out of the blue as I was watching TV at Lynn's house, the telephone rang. It was the jobcentre. I was told by a young man there that they had someone interested in the vacancy for a writer. The interested party was duly put on the phone and I asked him a few questions such as name, address and last job. Then it all came to light. He told me that his name was Granville Williams and he worked for the BBC at Pebble Mill, in radio. He told me that he lived on great wood road in Small Heath. He told me that he was interested in helping me to write my book but that there would always be the coughing involved. He was taking the piss basically thinking that I would be desperate enough to allow that little annoyance to be involved. I treated him courteously, taking his details down eagerly as evidence, and then I told him I would be in touch. As soon as I put the phone down I exclaimed aloud "yeah, as if". When I had said that, the man on the TV said "he's back!" So I guess that they must have been portraying me as a lost cause, seemingly allowing the BBC to take over the production of my book with harassment involved. When I fucked him off, that is when I was deemed to be "back" in contention again. This took place on Tuesday the 17th May 1994. Soon after that I had another row with Lynn again over her involvement in the conspiracy against me and we split up for a few days. We soon made up and spent the night together, after which she wrote me a note before she went to work the next morning declaring her love for me and outlining her disapproval of the telepathy being used as a weapon against me,

THE DARK SIDE OF THE BBC. A DISTURBING TRUE STORY.

though she does not use those exact words. she does however, refer to them as 'those bastards' which I found very amusing. The important pages of the letter are reproduced and kept on file in my office.

I saw my neighbours underneath, the Gameson's soon after from my kitchen window, and they were standing on the path outside with some tall chappie who was looking well moody, as if he was geared up for a fight. I was convinced that he was Lynn's ex husband (I had seen him once before) and I was quite happy to oblige, but by the time I got outside he'd conveniently gone, so I took a walk around the block to see if he would make an appearance, then came back into my flat...only to find that my wallet had gone missing along with my last thirty pounds inside it, (I'd left the door on the latch when I went out). There was no way I was letting the Gameson's and their counterpart play me for a sucker. I jumped on my bike and cycled down to Stetchford police station to lodge a complaint against the Gamesons and the BBC. I heard a girl's voice in my head say I was "mad", mad for going to the cops – let's face it, we've hardly been the best of allies - but I continued anyway, little knowing that because I now had another court case pending for criminal damage that would expose the BBC and their counterparts, there was another conspiracy brewing to keep me out of court by getting me sectioned under the mental health act as a deluded paranoid schizophrenic, again.. I told the officer at the front desk that I wished to make a complaint and waited patiently to be seen by someone. Sergeant McManus came down and before I could tell him of the theft, he informed me that he'd wanted to speak to me anyway: apparently my brother

THE DARK SIDE OF THE BBC. A DISTURBING TRUE STORY.

Mark was saying I'd burgled our mum's place. I knew what he was talking about straight away...I'd gone through the window of her house because I didn't have any keys, but I'd never stolen from the place. I used to live there. I was family for Gods sake. My brother Mark had obviously turned against me now! I'm not sure whether Sergeant Mcmanus was interested, because after I'd had my say and given him the beef on the Gameson's, the BBC and the Evening Mail he invited me to wait in a cell while he made some calls...my doctor, doctor Lee, a social worker and a psychiatrist duly arrived and enquired - from the other side of the cell door - whether I'd like to go to hospital. I knew what they were up to straight away so I went nuts, banging on the door and yelling at them for their involvement in a cover-up.

Obviously they somehow interpreted my tirade of expletives as a Yes, because they blatantly sectioned me and I was sent to Newbridge House Mental Health Unit on Hobmoor Road, Small Heath. They were trying desperately to keep me out of court and out of the limelight. The cover-up had begun. I took the opportunity to appeal against the section straight away, to an independent tribunal, using the representation of John Lloyd. In the meantime however, I was allowed to go back to my flat to sort out some clean clothing etc, escorted by two male nurses. When I arrived at my flat, I went into the kitchen by myself and after some self deliberation, I returned to the two nurses armed with a kitchen knife. I was still upset about being locked up in a mental hospital and decided that I would still get myself arrested for something more serious, even whilst I was Hospitalised, for a serious offence of kidnap and false imprisonment, just like had been done to me by the medical profession. I informed the two nurses of my intentions and decided to allow one of them to

THE DARK SIDE OF THE BBC. A DISTURBING TRUE STORY.

go free so that he could raise the alarm and thereby get things moving. I went to the front door with him and let him out of the flat. When I returned to the living room, to my surprise, the other nurse had escaped through a now open window. I nearly pissed myself laughing at the whole situation. I waited for about half an hour to see if a task force from the military would arrive, (the Police) but when they didn't, I took a slow walk up to my Mom's house in Yardley. The phone rung after about fifteen minutes. It was the Police. They wanted to know if I was there and my Mom told them I was. They landed there almost straight away (one of them was overdressed in full riot gear, including shield) and carted me off to Stechford Police Station. I was returned to Newbridge House the next day where I was hence forth treated as the enemy for some months after that.

Anyway, about my Solicitor, John Lloyd, I trusted him because he'd represented me previously when I was a 'foolish' voluntary patient in Hollymoor, via the magistrate's court. This time again he was clearly on my side so we decided to go ahead and proceed with our defence of a BBC conspiracy involving the existence and misuse of telepathy. The actual hearing would take some time though. In the meantime, I had an appointment to see my other solicitor handling the criminal court case of criminal damage that would expose the BBC. Throughout the interview he showed faith in me and made it plain that he would do the right thing and help me expose the BBC, But because he made it plain that he was on my side, he was taken over by the 'voices' in his head at the time to such an extent that he was forced to say the word 'hands' and he coughed repeatedly. Eventually, he came out with it and said "It's up to the tribunal". What he was saying now was that it was up to the tribunal if I

THE DARK SIDE OF THE BBC. A DISTURBING TRUE STORY.

should be deemed mentally ill and covered up or if I should be 'believed' and released from the shackles of a mental hospital to go on to expose the BBC in the criminal courts. I waited in the hospital for three months before the tribunal came. It didn't go as planned. The head of the tribunal asked, after hearing my story, whether the medical profession was involved in the conspiracy. I answered truthfully that it certainly looked like it, (stupid, stupid, stupid) and after that, my opposition, Dr. Eleanor Rollaston, then nastily interjected the words "six months!" They obviously responded by giving me the maximum stay of six months in Newbridge. In retrospect, maybe I should've said no. I was learning from each mistake. This was a high level cover-up

But I had a typewriter with me in the hospital and at least I had a guaranteed six months of no distractions in which to resume work on my book, which I was still very keen on at this stage. Lynn also visited me often during my stay. My Mom (God bless her) visited me almost every day. When my criminal damage court case came around I found myself being represented by a solicitor I'd never even seen before and who was against me from the off. He said something about "so *you* believe", when I mentioned the involvement of the BBC and I knew then that I was basically buggered, the expose was not going ahead . I knew it was going to fall apart and Lynn also agreed (because the case was now hopeless) that I may as well plead guilty, and do my time in hospital. The solicitor tried for a "not proven" verdict due to mental illness but the female magistrate judged me guilty and gave me a hospital order which piled up onto my criminal record. I was transferred to Small Heath Mental Health Unit on Chapman Road, where I was given anal injections once a fortnight until I went before the review panel of Doctors after the six months

THE DARK SIDE OF THE BBC. A DISTURBING TRUE STORY.

were up. The presiding Doctor, Doctor Peters, gave me another six. And then another six months after that. I spent a total of fifteen months consigned to Hospital, and even then I was only let out on licence, meaning I had to return to the hospital every night and sleep there.

Before I was let out on licence, I was downstairs sitting in the out-patients department one day when one of the nurses came up to me and told me not to panic but the BBC were outside the hospital doing some filming. I didn't panic. I took a slow walk through the hospital and made my way to the front doors where lo and behold, there they were in full force. It was early evening and it was a bit dark so they were using the big lights and all, but I just took no notice of them and sauntered back into the hospital where I made myself comfortable in an easy chair in the reception area. One of the so-called actors came advancing up towards me with a big plate of food he was munching into and a wicked grin on his face as he looked at me before veering off and out of sight. I later learned his name was Mark Porter And we would meet again some eight years later in another cover-up where I was held in Winson Green Prison in Birmingham. Eventually I was transferred to an outpatient, but my own flat had too much history for me to be able to stay there. My mum, who had visited me every day while I was in the hospital, let me move into her new house. I spent a year scrimping and saving my DSS money, I'd stopped drinking as well as smoking and had virtually no living expenses, so I accumulated over three and a half grand. I still came to the flat occasionally, if only to assert my ownership to Maxwell.

THE DARK SIDE OF THE BBC. A DISTURBING TRUE STORY.

THE DARK SIDE OF THE BBC. A DISTURBING TRUE STORY.

RING 'O' BELLS AND ROBBERY.

After a while though I was beginning to get bored just sitting in my mum's house and started going to the Ring 'O' Bells pub in Yardley some evenings. I timed it so I could go in, have a pint of mild and be at the house by the time my mum got back from the bingo. I liked it in the Ring 'O' Bells because no one there knew me, and I could just blend into the crowd. The telepathy had stopped temporarily before I was discharged from the Hospital and had still not reared its ugly head even now some many months later. I started playing pool there as well and through my table skills I got to know Roy Goulding, a regular there, another likeable rogue. We developed a real camaraderie and started going to other pubs together and it was then that I began to pick up again on signs that people knew me. Roy would always stick up for me though. Once we were in a group sitting around a table and the others started talking about me. Roy defended me, saying "He's as mad as I am!" They all laughed at the joke. Things took a turn for the worse though when the gaffer in the Ring 'O' Bells went on holiday, leaving a relief gaffer in charge. This guy was a bit of a divvy and used to give Roy free drinks because he was scared of him, but quite soon both of us and anyone else with us were getting lubricated for free, including whatever take-homes we wanted. After one session, when closing time had

THE DARK SIDE OF THE BBC. A DISTURBING TRUE STORY.

been and gone, we decided on a stopover. After several more drinks Roy rolled a joint, after which I swiftly blacked out. Roy, however, decided to take the piss and ask the divvy relief gaffer for some money...he obliged, handing over the entire contents of the safe, about three thousand pounds. 9am the next morning, Sunday, I was rudely awoken by the CID hammering the door of my mum's place where I was staying. Of course the person they really wanted was Roy. They assured me that if I told them where Roy lived, I'd be "out in an hour". I let them drive me around Acocks Green for about half that, telling them I didn't remember where he lived before they decided a few hours in the cells might aid my memory. I repeated my statement of ignorance when I was interviewed at five that evening, pointing out that I had been significantly inebriated the night before. When they asked if Roy lived in Acock's Green where they themselves told me he lived earlier, I answered yes, because I had taken them around that part of town where he was known to have been dropped off by the relief gaffer of the pub, after the alleged robbery. It didn't occur to me they might be setting me up, and this tape would be played to Roy to make him think I'd actually taken them to his girlfriend's place. When I eventually got out, I made a b-line for the Ring 'O' Bells the next day but Roy was nowhere to be found; his nephew was sitting outside with a mate, and he took one look at me, shook his head and turned away, like as if I was the enemy. I shifted to Roy's girlfriend's house after that to find out what was going on but she wasn't in. It was only a couple of days later when I managed to catch her in that she told me the word was I'd grassed on Roy. I defended myself but it was useless. Obviously I was now public enemy number one for Roy and his family. For several days I saw neither hide nor hair of Roy. Terry

THE DARK SIDE OF THE BBC. A DISTURBING TRUE STORY.

Gardner, one of our drinking friends in the Ring 'O' Bells said he'd seen Roy in there with some other guy looking real moody, and that I should find him and tell him straight what had really happened, how the Police had stitched me up in order to cause trouble for me. I waited in there for an hour but he didn't turn up. Several days passed with no word and in the interim my mind turned to other things. Like the BBC controlling Birmingham against me. Their telepathy had slowly begun again but I was ready for them this time. I knew I had to get myself arrested and turn the case against them at an actual trial now. But it would have to be a crime too serious to be dropped, something which couldn't just be swept under the carpet. The voices in my head were telling me to murder someone involved in the conspiracy against me, but armed robbery seemed a better idea. I made some enquiries to a trainee solicitor called Chris , who worked part-time as a barman at the Ring 'O' Bells and found that I would still be charged with armed robbery if I only passed a note to the bank teller saying I had a weapon. I didn't want to mess around getting hold of a gun (I don't like dealing with those things).

On a Monday, in the early part of 1998 or 1999, I sent off a brief version of the book, along with evidence of the conspiracy against me, to Ian Gold Solicitors. This would arrive the next day, Tuesday. I was now ready to commit my crime that would enable me to expose the BBC. At about 1:40pm I caught the number 17 bus into Small Heath, thought broadcasting all the time. I was going to hold up the Nationwide Building Society on Coventry road, Small Heath. I wasn't going to go steaming in there straight away: I figured if I waited around a little I'd have less time to spend in the cells. 2pm ; I went into a cafe opposite the intended crime scene, sat down and had a drink and spent some

THE DARK SIDE OF THE BBC. A DISTURBING TRUE STORY.

time contemplating. The telepathy was strong, and the staff were reacting to my thoughts as I sat there. 4:30 I went and bought a paper and a book to read in the cell, and got a bag to conceal my so called 'gun'. I went back into the cafe for one last coffee. The waitress there smiled at me and said "I thought you were going to rob a bank or something!". I smiled back and told her "No bab, not yet", then drank my coffee in quiet telepathic concentration before leaving to check out the joint. 4:50 I was in the queue at the Nationwide Building Society. 4:55 it was my turn. I ambled up and passed the female teller the note I had pre-written which said, "I have a gun, give me the money". I had noticed signs saying 'Do Not Lean over the Counter', so I knew they had concealed bullet-proof barriers which would shoot up at the touch of a button, which they did as soon as the girl saw what my note said. She ran off to phone the police as I stood and waited. 4:57 and the cops walked in, smiling. "Come on then Tony" they said, and walked me out. They obviously knew what I was doing before they were called. There was a crowd outside and I heard an Asian kid say in mock-alarm, "He's tried to rob a building society!". The arresting Police Officer called him a "wanker" and told him to "get lost". So I was back in Stetchford Police Station, and they were treating the whole situation with respect this time. They even loosened the cuffs for me when I was in the van. They charged me with attempted robbery, and when I asked why I wasn't charged with 'armed robbery' the Police replied "This ain't the bill you know". In the cells there were no serious voices, unlike my previous experiences, just a sense of calm, awaiting the inevitable, my day of victory. A WPC came to ask if I wanted anyone informed (as if everyone who mattered didn't already know) and I said just tell my mum. However it turned out even she knew already. Unbeknownst to me,

THE DARK SIDE OF THE BBC. A DISTURBING TRUE STORY.

my sister Helen was there at the house when the Police arrived to tell my mum, and took it upon herself to attend the Police Station in order to stultify my progress by letting all and sundry know about my false mental health history. The BBC had turned all my family against me by now. It was then that things took a turn for the worse. Within half an hour the Police had started clattering my head with 'voices' and in an hour the cops were requesting my presence in an interview room where I found two doctors waiting for me. Oh shit! Not another cover-up! They enquired my opinion regarding the time, date, and who the prime minister was. Their approach made me laugh though, and I went through the interview with them, cracking jokes, and they loved me. When they told me about what my sister was saying, I had to think fast so I just replied that we don't quite see eye to eye and she'd happily say that to get me into more trouble. At this point a copper walked in to enquire if I was to be sectioned under the mental health act but one of the Asian Doctors jumped up and proclaimed rather fiercely "you must charge him!" To this, the copper retreated rather sheepishly. Then another one appeared to tell me my sister was outside, asking to be present at the Police interview; I told them she could not. There was no way I was letting her in after what she'd said about me. I then went on to tell my solicitor, a nice looking legal clerk from Ian Gold's office, about my plan, why I tried to rob the Nationwide, making sure I explained it was Carlton TV I was seeking justice against, not the BBC. Obviously this was the right thing to say, because one of the cops came into the interview room and said "Yeah go that way". He was under the impression that I thought it was Carlton TV that was responsible for trying to make my life a misery with the use of telepathy, not the BBC. He thought I was going to be foolish enough to only blame Carlton TV

THE DARK SIDE OF THE BBC. A DISTURBING TRUE STORY.

and let the BBC out of the picture. I opted to make a no comment statement, but the Police interviewers were cunning. Usually if you make a no-comment interview, none of the transcript can be used in court; but if you answer even one question, then the whole thing becomes admissible evidence. In this case I was doing fine until they threw in some trick question and I automatically said "No", thereby invalidating my careful attempts. But it made no difference anyway as I continued to say "no comment". I was charged with attempted robbery, refused police bail and spent a cosy night in the cells. The next morning I was taken to Birmingham Magistrates Court, where I was greeted by a black girl from social services who asked if I had any previous serious convictions. I told her no, I had none. I was facing a serious charge now, as expected, so the possibility of bail was uncertain, but when I appeared in the actual dock before the Magistrate, my solicitor and hero Ian Gold exclaimed rather snottily, "I think I can get you bail, but you really must see a Doctor!". What a bastard! He then went on to tell the Magistrate of how I should be taken to a mental hospital and given "injections", sneering at me all the time as he spoke. To this the Magistrate leaned forward and enquired "so we are agreed on the way forward?" They were blatantly conspiring together against me. I wasn't agreeing to this at all because I knew straight away that I had been betrayed and was being swept under the carpet. I was given bail anyway due to the fact I apparently had no previous serious convictions, but unfortunately for me, Ian Gold had implemented another big medical cover-up.

So I found myself unceremoniously escorted to the street outside where I paused to clear my head and think through my options. The most pressing was my legal representation: I didn't feel happy with Ian Gold after his

THE DARK SIDE OF THE BBC. A DISTURBING TRUE STORY.

attempts to bring the medical profession into the case, so I needed someone else from the same firm, but someone more sympathetic to my predicament. As I thought this, I saw the people in the distance all raise their heads as if they had heard me think this. A woman walked past me and said "Yeah do that". Thus reassured, I walked down to The Queens Arms pub on Steelhouse Lane and partook of a few bevvies to loosen me up. I was back at my mum's by 3:30 pm when she returned from the bingo. Admittedly I was quite angry at my sister Helen's previous antics and after a few harsh words, my mum went round to my brother Mark's. I had no idea he'd phoned Doctor Harrison until the good Doctor turned up at my flat (where I was stopping occasionally by now) the next morning with Indira Kattimani, the senior social worker. They repeated what Mark had told them: apparently I had "gone berserk", "threatened" my mum and "smashed the house up". I had done no such thing and I told them so. I advised them to phone my mum and check out Mark's story but they were unwilling to do so; instead they left to phone Mark again. They were true to their word, unfortunately. He told them about the robbery, and my plan to expose the BBC. Imagine my surprise when a whole bloody army of conspirators landed at my front door 5pm the next day, a Saturday. There was a social worker and a second doctor (as required to carry out a mental health section), both female and two coppers, one male and one female. I clocked them through the spy-hole in my front door and knew immediately what was going on. I ran back into the living room and dived out the window (one storey up), stopping only briefly to turn the music off on my way out. I landed ungracefully on my arse and before I could even gather my bearings, one of the cops was standing over me. "What are you doing?" He enquired cordially. I said I thought they were going to arrest

THE DARK SIDE OF THE BBC. A DISTURBING TRUE STORY.

me, but he assured me they just wanted a word. There wasn't much point me running away now, so I took them all and their smiles up to the flat. The first thing the second doctor said inside was a comment about "them people" not usually being so tidy. She was referring to mad people. My flat was spotless, as always, and for some reason this didn't gel with her preconceptions...if I was mad, I should also be scummy, obviously it hadn't occurred to her that I might be neither! Her observation put me in a good mood though and of course when the old Hickey charisma kicks in, people can't help but react to it. The social worker (a pretty black woman) was standing in the background analysing me, but I could tell she was loving it because I had tried to escape from them out of the window, but the doctor wasn't quite so positive. She was of the opinion that a stay in the hospital would do me good (cover-up, cover-up). I responded alarmedly in the negative, that I did not need to go to hospital and then the social worker piped up in my defence for the first time since coming in, saying I didn't need to go to hospital. She liked me! There was a moment of silence and glances between the medical bods then the female doctor (a not so pretty Asian woman) said they needed to talk in private. They went into my bedroom while I kept the police company. Of course, I was thought broadcasting all the time to everyone outside as this was big news and people were kept aware of the course of the conspiracy. We discussed the robbery charge and the sexy WPC said in a more raised voice "If Tony had a gun, he would have used it". She was letting people know that I should not be treated badly in this way as I could be dangerous. I wouldn't really have used a gun at this time, but at least she was trying to help me. After a little more chatting the medical bods came back in, grinning like mad and told me what I wanted to hear, that they weren't going to

THE DARK SIDE OF THE BBC. A DISTURBING TRUE STORY.

send me to hospital, at which point the doctor mumbled into the microphones "Because he's too clever". So the merry party left and I was buzzing my tits off. I had to do something so I made my way over to my mum's. An uneventful journey except for an Evening Mail guy who stared with immense hatred for me at the floor as I walked past. He was unhappy for me to have escaped the big medical cover-up. I considered attacking him, but knew that if I did I'd definitely be sectioned. My mum wasn't around, probably because Mark had told her to stay away in case I returned. I had the keys anyway, so I let myself in and stayed the night there. Sunday, and still no-one had turned up. I decided as the court case was definitely on the roll I should start writing again. I began from the day of the attempted robbery but I kept being interrupted by strange, silent phone calls. Finally someone talked to me, and it turned out, this time at least, it was Roy Goulding. I was surprised to hear from him, this little incident having slipped from my mind in the melee of the last few days. I told him the truth about what had happened and we agreed to meet in Yardley. He never turned up, but I found him outside my mum's house a few days later where he glared at me and said "So there he is, the man I can never seem to find". I went through what had happened with him again but instead of believing me he got aggro, which just flared me up even more...when he saw I wasn't frightened he calmed down himself and we took a walk up to the park and he accepted that I hadn't grassed him until a few days later when, probably under the Evening Mail or BBC's influence, he started to doubt me again. He wanted to hear the police interview tape and of course I got it for him, planning for us to listen to it together. Unfortunately the way things worked out, we got pissed and went our own separate ways. When Roy heard the set-up confession, the bit where I

THE DARK SIDE OF THE BBC. A DISTURBING TRUE STORY.

'admitted' showing them where he lived, he can't have been impressed...anyway, I didn't hear from him for quite a while after that and I had returned to my task of compiling my story. The silent phone calls continued, and once someone asked to speak to me. I was more than a little suspicious by now, so I asked who was calling and he said he was from the probation service. Liar, I recognised his voice as the Irish copper who interviewed me over the Ring 'O' Bells incident with Roy. I told him my brother Tony wasn't in and he said he'd call back later. I had to pretend to be Mark. But before later had a chance to arrive, an ambulance turned up at the house. They rang the bell and I opened the door looking perplexed. They also were looking for Tony Hickey, but as I told the copper, he wasn't around. I eyed them carefully and asked why they wanted him. Turned out they had instructions to take him to Highcroft hospital and were to be met here by a Police escort. I laughed quickly and shook my head, "I'm Mark, my brother Tony is upstairs, asleep. He doesn't need to go to hospital. You've been wound up, mate". And I knew who had done the winding...it was that Irish copper again trying to get me put in a mental hospital again in order to cover-up the impending Court Case. But my brain had outdone his clockwork machinations. I watched through the window as the Ambulance driver got back in and picked up his radio. He sat talking for ten minutes then drove off. Of course by now I was proper unsettled, and expecting another visit from someone looking to curtail my freedom. I dodged out of the house and hid in the park across the road, from where I could keep an eye on the front door. After an hour of this I had calmed down a little and went back into the house to continue writing. I hit the sack finally at 2am Monday night. Tuesday was Giro day for me and I was up with the birds at 7:30am. I was about to leave the house at 8am

THE DARK SIDE OF THE BBC. A DISTURBING TRUE STORY.

when there was a knock at the door. Eight coppers, Indira Kattimani, a senior social worker and my brother Mark greeted me. None of them looked particularly happy. I asked what was going on and the senior social worker involved told me I was coming with them. "What do you mean I'm coming with you?" "You're going to Highcroft Hospital" he replied. At this, I slammed the door, unaware that one of the cops had his foot at the ready, blocking the way. Of course I couldn't run because of my injured foot I got from jumping out of my flat window previously. The coppers jumped on me, cuffed me and dragged me into the car. I was driven to the hospital and when we arrived I was left waiting in the Police car outside the door to the intensive care unit. All the patients inside there were watching me, in awe of me now that they could see me in real life, as my subsequent ensnarement had been broadcast throughout Birmingham.

THE DARK SIDE OF THE BBC. A DISTURBING TRUE STORY.

HIGHCROFT HOSPITAL.

I was taken through the Dining Room and a girl said out loud "He's alright", but this didn't convince any of the officials. We arrived at a bedroom and they told me this was where I was staying. I had some breakfast in the dining room, then I was taken to a room to see the social workers involved in getting me sectioned. One of them was a male and the other one was Indira Kattimani, an Asian female. I asked why I'd been sectioned, and he pointed out that I was intelligent enough to know why. I receded, knowing there was no point raving about it. I was here now and I just had to ride it out. My previous experiences in hospitals of this sort had prepared me. I knew the score, I knew I had to waste no time in putting in an appeal. This would take three months anyway to traverse the beaurocracies of the medical profession. The days and the weeks passed by. All the inmates knew who I was and there was a distinct renaissance of telepathy in

THE DARK SIDE OF THE BBC. A DISTURBING TRUE STORY.

there, but I was still buzzing about the court case. Everyone was cool to me and this was probably my most pleasant period in the hospital system. The only thing that got me down was seeing the consultant psychiatrist every week. This was a very nasty piece of work called Doctor Jon Kennedy and he was constantly probing me and questioning me, trying to prove me insane. I denied any association with the BBC. This was all in the past, I told him. Unfortunately I didn't know that my brother Mark had found my book I was writing at my mum's place and told the doctors all about it. I also denied any knowledge of this when I was questioned on it. I was steadfast in my ostentatious belief that I'd been sectioned because of Mark saying I'd smashed up my mum's place, which just wasn't true. My mum was still coming to see me though, being driven to the hospital by my brother Michael. Michael was the most sympathetic of my brothers and when I asked him to get the book off of Mark he obliged without question. I commenced writing again although I wasn't going to tell anyone about it. In fact I had very little communication with the other patients except for Dawn Sturdy, the girl who had spotted me in the dining room on my first day and declared me to be "alright".

So I was spending almost all of my time writing, building my expose against the forces of television. The officials knew what I was doing, but I just told them I was writing letters. I spent about three months solidly working on my book before my paranoia got the better of me: if anyone did find the book, I'd be in big trouble, and Doctor Kennedy himself was always asking me about it as he could then 'prove' it was evidence of my insanity. The next time Michael brought my mum to visit I gave all the work I'd done on the book back to him to take out of the Hospital. When he asked if he could read it I said yes; I trusted him

THE DARK SIDE OF THE BBC. A DISTURBING TRUE STORY.

because he'd helped me out in bringing the book to me in the first place. But my trust was misplaced. On the morning of my first appeal against being sectioned to the Hospital Managers, a process involving a committee of three doctors, I was feeling quite positive about the prospect of freedom. I was discussing my case with my solicitor, Jon Lloyd, when a knock on the door told us someone wanted to see me. It was my brother Michael. He walked in and announced that he'd given my book to Doctor Kennedy. I went berserk. Apparently, he told me that some of my family had had a meeting and decided this was the best thing to do. I couldn't believe it, but I had to calm down quick before the appeal, even though I now knew it was all hopeless. The medical conspirators now had the false 'evidence' they longed for to keep me locked away in Highcroft Hospital, indefinately. The cover-up was in full flow! When the appeal finally got started after hours of waiting, my solicitor left after the first fifteen minutes because he was late for an 'independent' tribunal at another hospital and it was then that the hospital officials, the so called Hospital Managers, dropped their pretence of respect for me. I sat and took it for a few minutes more before walking out, inviting them to try a variety of unusual sexual positions as I did so. After I left a nurse told me Michael was really worried that I hated him and I should go and speak to him. I didn't particularly want to, especially after what I'd just been through but it seemed like the right thing to do. He was family after all, even if he wasn't a fighter like myself. I was so used to being let down by my family now that making peace with him wasn't that hard.

So I had three months to wait before my next chance of freedom. The next appeal I had coming was the Independent Tribunal. This was an outside tribunal that had no connections with the Hospital itself so they were less

THE DARK SIDE OF THE BBC. A DISTURBING TRUE STORY.

likely to partake in a criminal conspiracy involving a medical cover-up. So I believed. This tribunal also consisted of three people, one from the medical profession, one from the legal profession and one lay person. Again, my solicitor in charge of the case would be Jon Lloyd. What had happened at the Hospital Managers meeting was public knowledge and one of the staff there, a big fat black guy, had started taking the piss, cracking jokes about me being a writer, a softy...until one of the patients (a youngish black guy also named Tony) pointed out to him that I wasn't just a writer: I was a fighter as well. That shut him up, and he didn't make light of my talents after that.

There was sporadic bursts of telepathy involving all the other patients being 'controlled' for the next few months while I waited for the independent Tribunal to come around, but mostly the invasion was confined to the 'voices' in my head now and from the television in the communal room. On one occasion when all the other patients were being controlled by telepathy in the continuing bid to harass me, one of the few girls there, Dawn (the one that declared "he's alright"), came up to one of the male nurses in front of me and said to him out loud "I keep hearing voices in my head telling me what to do". She was letting me know that she wasn't responsible for her part of the harassment I was receiving as we had often spoke openly about the conspiracy and the use of telepathy that went with it. Dawn was a pretty little thing of about twenty years old. She wasn't mad like the rest of them, she just had a self-harm complex which I told her was a bit daft really to say the least. We became close over a short period of time and eventually we became sweet on one another, sharing a few romantic clinches in the process. We were going to go the whole way one night but decided against it at the last moment. It was not the right place for

THE DARK SIDE OF THE BBC. A DISTURBING TRUE STORY.

either of us really. She was transferred to another Hospital in Northampton eventually but we kept in touch all the time by telephone and by letter. It was nice to talk with someone else who wasn't off their head.

Before the independent tribunal took place, my solicitor, Jon Lloyd decided to get an independent assessment of me done by another psychiatrist which we could use at the tribunal, if it proved worthy. We employed the services of Dr. Van Woerkem who Jon had used on previous occasions and who later came to assess me at Highcroft Hospital, where I was still incarcerated against my will. He was well aware of the conspiracy I could tell, but he took a shine to me and wanted to help me in my bid for freedom. He even told me that there was a document in my file which read 'not to be shown to the patient'. He smiled broadly and his eyes lit up like beacons as he read through it to himself so I gathered that it was conspiracy orientated. I was going to snatch it out of his hands as he read it but decided against it in the end as I didn't want him to turn against me. I told him that I believed that I had been sectioned under the mental health act because of my brother Mark's lies about me and because my book had been found, I also had to admit to my belief that there was telepathy in operation in Birmingham. . He wrote a decent report about me anyway but it was all to no avail. The independent tribunal were as bad as the rest of them and my detention in Hospital under the mental health act was to stand. I was gutted but remained positive. It dawned on me then that I wasn't just up against the dark side of BBC television, but that I was up against the whole rotten establishment. I continued to write my book openly now without fear of detection, due to the fact that it was no secret anymore and no-one in authority believed that I would ever get it published now anyway.

THE DARK SIDE OF THE BBC. A DISTURBING TRUE STORY.

I wasn't too surprised when the six monthly review came around to determine whether I should be detained further in hospital and it resulted in my being sectioned for another six months. Kennedy had it in for me and no mistake. I wasn't too surprised, but I was still angry. I just called him a bastard to his face before storming out of the room this time, and then tried my hardest to kick in one of the windows in the communal room so that I might make my escape. I failed, and limped back to the room to ask Kennedy why I was being kept in. "We think you're a paranoid schizophrenic," he answered. "Sweet, whatever" I told him. It was only another six months, the same as I'd already spent there. I got another appeal in straight away. But then I was transferred to the long-term ward, a worrying new development. They were trying their damnedest to keep me out of the limelight, to prevent me making official what I knew. And there was jack-all I could do about it on the inside. To keep my mind off the situation I started training, going to the gym everyday and keeping myself in shape.

As I was getting more and more into my weights and boxing, (they actually had a punch-bag in the gym there), I noticed the telepathy that controlled all the people around me and the 'voices' in my head started decreasing. But it wasn't until after my next court appearance, which was at Birmingham Crown Court, that the use of telepathy by the dark side of television stopped completely. I had been due to appear in court the previous month in relation to the charge of attempted robbery of the Nationwide Building Society, but for no true reason at all, apart from the fact that Doctor Kennedy was now leading the conspiracy against me and trying to pervert the course of justice in order to cover-up for the illegal activities of the BBC, he wrote a letter to the

THE DARK SIDE OF THE BBC. A DISTURBING TRUE STORY.

courts saying that I was too ill to attend court on this occasion due to serious mental illness and that my hearing should be remanded in my absence until such a time that I was well enough to attend court. Absolute bullshit! I was probably the most physically fittest, mentally strongest individual in the whole of the entire Hospital complex. I didn't even give a damn about the use of telepathy against me and I feared no-one. In reality, what Kennedy had done was to make it 'official' to the courts that I was mentally ill so that the criminal conspiracy against me and the cover-up of the illegal activities of the BBC could now take place within our judicial system. He was in fact sabotaging my Crown Court trial. Although I cannot prove it to be true, he was probably in collaboration with my own so-called solicitor and false hero, Ian Gold. Doctor Kennedy and co, were slowly trying to demoralise me. Aiding and abetting my ruination.

With all that going on in the background, I was escorted to Birmingham Crown Courts in a taxi by two male nurses for my final appearance in court before the actual trial. Although I was on conditional bail of residence to my Mothers house, curfew between 10pm and 8am and had to stay away from the Nationwide Building Society, I was still a free man in the eyes of the law because I was not remanded in custody to await my trial. But I was taken to the lock-up's underneath the courts and kept under lock and key until my appearance before the judge. When I protested to the little Welsh nurse who had apparently assumed command (his name was Idris), that I was not supposed to be held in custody and that I should be allowed to walk into court through the front doors like any other innocent man, I was told it was a technical thing and that it was standard procedure. This was just more bullshit because I had been taken to court before from Newbridge House remember? But I was not held in custody in

THE DARK SIDE OF THE BBC. A DISTURBING TRUE STORY.

the lock-up's on that occasion. This was all designed to make me look unacceptable in court. The existence of this book was now a real problem to the conspiracy. Anyway there was nothing I could do about it at all so I just had to bite the bullet yet again. Eventually, when I did make my way up the steps to the court room where I was due to appear, I was met by the Barrister that Ian Gold had appointed to represent me in the Crown Court. We talked very briefly together in a small room outside the court in the presence of the two male nurses. I was given no privacy to discuss my case with my Barrister. In fact I wasn't even given the chance to discuss my case at all. It appeared obvious that I was to have no say in the matter whatsoever. My Barrister inquired as to whether I was happy with my stay in Hospital, but when I began to show my concern about being hospitalised, he cut me off discourteously and sneered "well you're there now aren't you?"

Then, just to make the cover-up more official, he turned to the little Welsh nurse Idris and said "shall we get him remanded to Highcroft Hospital?" To which the Welsh one beamed with pleasure and replied "yes". This made him feel more important than he actually was so he was dead chuffed to be involved in something so big. T he proceedings in court whizzed by after that as I was in a state of total confusion. All the months I had spent locked up in a mental Hospital, pinning all my hopes on this court case (there can be no whitewash at the whitehouse!) were now seemingly wasted on nothing. I had been betrayed yet again and it was slowly but surely sinking in. There was not going to be an expose. I pleaded not guilty and was remanded back to hospital to await my trial.

THE DARK SIDE OF THE BBC. A DISTURBING TRUE STORY.

It took some time to actually come to terms with this cover up at the Crown Courts, but it finally dawned on me some days later when I was back in Highcroft Hospital. I was re-living the events in my mind that took place that day outside the court-room with my Barrister and when I remembered him saying "well you're there now aren't you?" It was then that I realised it was all over. At that point, a 'voice' in my head who I know to be one of the more understanding than some of the rabble said "that'll do!" and then the voices gradually faded away to nothing. Soon, there was no invasion of my mind, no other presence's grating at my consciousness and I reached a plateau, a state of calm lucidity. I was glad to be rid of it, but I was not convinced that now the battle finally was over.

Three months passed, and it was independent tribunal time again. The story I designed for this appeal was only a half lie: I told them that although I knew I had been ill, I was now fully recovered completely with it and free of my past delusions. My solicitor, John Lloyd, did real good putting my case across with the help of an excellent independent psychiatric report from Dr. Van Woerkem who backed up my claim that I was no longer mentally ill. Everyone present at the tribunal hearing was well aware of the use of telepathy and the fact that I was in Hospital to cover-up for the use of telepathy so it was not a foregone conclusion that I would win my appeal against being sectioned under the mental health act and be allowed to go free. My solicitor, Jon Lloyd also knew this and so while he was saying his piece to the tribunal members, he interjected fiercely with the words "how long?" I remember at the time that this gave me some immediate cause for hope because it was plain to me, and all the others what he was saying. He was by-passing the issue of whether or not I had

THE DARK SIDE OF THE BBC. A DISTURBING TRUE STORY.

a mental illness, let's face it, they all knew I was not insane, but now in a blatant counter conspiracy, he was asking the tribunal "how long?" do they intend to keep me locked up illegally in Hospital. I had been hospitalised for nine months now, which was outrageous in itself, but it was also common knowledge that the court case was well and truly going to be covered up as well. I could see the concern on the faces of the tribunal members when Jon said this to them as it put the fear of God into them to think that someone in authority was not going with the flow of the conspiracy which could result in they themselves being exposed for involvement. When the hearing duly ended and the tribunal retired to reach a decision, I was asked to leave the room while they deliberated, along with my solicitor Jon Lloyd. We talked positively outside about the possible outcome of the case and after thirty minutes of private discussion the committee called me back in and informed me I was free to go. I had been successful in my appeal against being sectioned! And we all know why don't we? Thank you Jon!

I was free to go! After nine months being locked indoors with a bunch of psychos, I was let out into the real world again! The nurses tried to tell me to stay that night at the hospital and leave in the morning after I had said goodbye to Doctor Kennedy. Were they taking the piss or what? I knew when I was leaving, and it wasn't the next morning. I walked out without even stopping for my things. I headed for my mum's house but she wasn't at home so I made a b-line for the Ring 'O' Bells. I hadn't had a drop of alcohol for three quarters of a year and, to put it bluntly, you could soon tell. I had four pints in the Bells and found myself face down on the ground several times during the walk back to my flat. The next day I tidied up my loose ends, collecting my possessions from Highcroft and giving them one final V as I left.

THE DARK SIDE OF THE BBC. A DISTURBING TRUE STORY.

To date now, I had been locked up in mental hospitals on three separate occasions. The first occasion was for a period of three months, the second was for a period of fifteen months and the third was for a period of nine months. All the people involved with my illegal incarceration were well aware of the existence and misuse of telepathy by the BBC and most of them, if not all, used their positions of power and influence to initiate a cover-up for the illegal activities of the BBC , thus allowing them to continue to randomly select and abuse other innocent, ordinary members of our general public. Makes you think doesn't it?

THE DARK SIDE OF THE BBC. A DISTURBING TRUE STORY.

CROWN COURT TRIAL.

I had less than a month until my court case for holding up the Nationwide. My nephew Shaun and his pals had been staying in my flat since he was released from prison a couple of months ago, and they'd turned it into a right shit hole. The second night after I got out of Highcroft I kicked him out of my bedroom onto the couch, and although this meant I got some peace at night, it was still not ideal and I wasn't at my best for my appearance in court, even though I knew it was all going to be covered up. I wasn't going to say anything about the BBC or the Evening Mail to Shaun at this time. They'd left me alone for a good

THE DARK SIDE OF THE BBC. A DISTURBING TRUE STORY.

few months now and I didn't want to rake up the past. When I eventually went to the Crown Court for the actual trial, I had already pleaded not guilty and the prosecution went quite easy on me to begin with. When the male prosecutor tried to say I intended to actually rob the Nationwide, I told him no, if I had tried I would have succeeded, and I definitely wouldn't have stood there and waited for the cops to arrive. I told him that I only committed the alleged offence in order to get myself arrested and expose in court at a trial, the existence of telepathy against me in Birmingham because I was mentally ill as my barrister had said. I knew things were going to work out when I heard the judge doing her summing up at the end of the trial. The existence of telepathy may have been covered up due to the medical conspiracy, so I just wanted a not guilty verdict now so that I could just get on with my life as best as I could. The judge herself almost indicated to the jury that I should be found not guilty, so she was safe, she took my side as best as was possible and the jury came back with a not guilty verdict. My ordeal was over at last. I wouldn't be going to jail or back to Hospital. Possibly the strangest period of my life ensued after the court case. All remained quiet on the telepathy front, but I wasn't myself. Shaun and his new girlfriend Kerry Quinn were still dossing down in my living room and they were not really paying the respect due to me for letting them stay in my home. They were both heroin addicts at this time and it was getting me down. So much so that I spent almost the whole of the year 2000 in one long stifled depressed mood.

By the time Christmas came around I'd completely forgotten that the BBC ever had anything to do with me. I was so wrapped up in my day to day life that the

THE DARK SIDE OF THE BBC. A DISTURBING TRUE STORY.

whole conspiracy had dropped out of my awareness. Robert Maxwell was out of sight, out of mind. Until Christmas.

Christmas 2000, my sister invited me to the club her husband now managed at that time, the 'Sedgmere Sports and Social Club'. They have a Christmas dinner every year there, invitation only. I was one of the elite few, one of only about thirteen people invited. It seemed like a good idea, a chance to get out and socialise, do something different. Shaun and his ex-girlfriend Hayley had also been invited, and my nephew and I arrived together, greeting my sister and Hayley in the hallway. We walked through into the bar to get something to drink. I noticed Maxwell as I walked by the bar. He was talking on a mobile phone and had a concentrated, stern look on his face because he was watching me in his peripheral vision. It didn't make any real connection to my past at the time. He was just someone I'd known from a few years ago who I had had suspicions about, the whole conspiracy was just a distant memory to me now. One that I was happy to forget. The other three trapped off somewhere rather suddenly and I went into the lounge and bought myself a pint of lager. Then my sister Helen, my nephew Shaun and his ex girlfriend Hayley arrived and immediately decided to go out into the hallway. Hayley told me to come out into the hallway with them, which I did in all innocence, little knowing that I was being set-up for a meeting with Maxwell to see what my reaction would be after all this time. Maxwell soon followed us out. He came up to me, still looking rather concerned and started talking to me about a mutual friend, Paul Kenna. We chatted amiably for a very short time, then when he was sure I was now unaware of his involvement as the evil one recruited by the BBC or their counterparts, he suddenly came alive, shook my hand profusely and walked away into the bar

THE DARK SIDE OF THE BBC. A DISTURBING TRUE STORY.

area like he was king kong. It concerned me that he'd shook my hand in that way. It concerned me the way he looked constantly at his mobile phone without ever once looking into my eyes, his manner, his every action concerned me. We hardly knew each other, after all. We were barely acquaintances. He was hiding something. There was nothing I could do anyway as I was not going to start a fight in my sisters pub at Christmas. I would look like the bad guy then.

We all went into the dining room to have dinner, and as we were eating I felt an old, familiar sensation. The BBC were communicating with the other guests. I could tell by their reactions and the things that were being said in front of me. Their old tricks had begun again because of Maxwell's great escape which must have left me looking a bit foolish. In order to counteract this, the telepathic broadcasters that were concerned about me told the rest of the dinner party "Tony's going to kill someone soon". They told people this in order to keep them wary of me and not brave enough to try to take advantage of me. This was only half an untruth as I had thought about killing someone on a number of occasions, who else in my predicament wouldn't have? But I certainly wasn't going to kill anyone over Christmas dinner at my sister's place. That was just ludicrous; I was too busy enjoying myself. Then when one of the guests there, Christine Stokes exclaimed aloud, "that's just rubbish", I eventually put two and two together after a couple of days had elapsed. I sat through the dinner, mingled a bit afterwards then went back to the flat. It was the next day when the voices began properly. Maxwell had realised I'd forgotten all about him, and taken his opportunity to attack me again. Some of the other dickheads followed suit, and quickly my mind; my sanity became a joke for them once again. Everything came flooding back to me; I remembered what I'd forgotten for so

THE DARK SIDE OF THE BBC. A DISTURBING TRUE STORY.

long. I remembered Maxwell and his cowardly abuse and I remembered his usual catchphrase of "look at him!" Said into my mind with as much nastiness as he could muster. He continued to use that same nasty catchphrase again now as he thought he was in the clear again so I knew it was him that had been involved all along. I wasn't willing to go back to the way things were two years ago. But a plan came to my mind.

I had found out about something called Disability Living Allowance in Highcroft. Most of the other patients were on it and it meant that they got between £28 and £90 a week on top of all their usual benefits. I actually spoke to Doctor Harrison (the same one that got me sectioned the last time) and he assured me I would be able to get it backdated to the date I left hospital, which according to my calculations meant I'd have more than a grand coming, assuming I got the lower rate of £28pw. Around this assumption, I made up my mind to go to Australia. The BBC and the Evening Mail wouldn't be able to reach me there. I'd be safe for good, free of their telepathy, free of Robert Maxwell and his attempts to destroy me. In other ways life was improving. In January Shaun got banged up again for driving while disqualified. I put all his stuff out to pasture in the shed and redecorated the flat in white, got a telephone, a brand new washing machine, a brand new tumble drier and hey...suddenly life is looking good again! Shaun and Kerry are out of my flat, things are looking nice, and I'll soon be off to Australia to sanctuary. I went to the Citizen's Advice Bureau to get a second opinion about Disability Living Allowance. It was Kathleen Corrigan who saw me and she was adamant that I could only claim this allowance from when I first send in the forms. This was bad news. I was expecting over a grand and when the claim was eventually processed the cheque

THE DARK SIDE OF THE BBC. A DISTURBING TRUE STORY.

I received was for a measly £312.40. This was on 14th February 2001. I realised that if I was going to go to Australia I'd have to do some serious saving, and it would probably take me over a year to get the money together. In the meantime I had some immediate expenses to outlay: I went out and bought a punch bag so I could start training again. I decided to cut out the fags and the booze, commit myself to a year of scrimping and saving, and if I was strict with myself I could save £60 each week and have the money to escape down under. It didn't quite work out that way. I had bumped into my old mates Peter Flynn and Dessie Mulcahey and started going out revelling with them again, which was proving to be quite a drain on my finances. Australia was looking increasingly distant, but I wasn't willing to let things get back to how they used to be. Maxwell had pushed his luck too far this time. If I couldn't escape, I was going to have to stand and fight them yet again. I had to resort to my old plan: an armed robbery. I was going to go out, get arrested, and turn the court case against the conspirators. The good voices, the ones on my side, agreed that this was the best course of action. I would definitely not be using the services of my last solicitor, Ian Gold. That would be a dreadful mistake. But I had to plan carefully. I couldn't risk being sectioned again. I needed to take some time and plan exactly what I was going to do, how to make them regret ever coming near the inner privacy of my mind. This time I would plan it meticulously, make no mistakes, and see it through to the courts where I could expose the BBC and their crimes against me.

THE DARK SIDE OF THE BBC. A DISTURBING TRUE STORY.

TRY, TRY, TRY AGAIN.

THE DARK SIDE OF THE BBC. A DISTURBING TRUE STORY.

I started checking out the building societies in the city centre. But as I was doing so I realised that the BBC were broadcasting my thoughts, telling everyone around me what I was planning. Some of them were on my side – I heard several people say "He will" and knew that they were trying to reassure me in the face of the looks and comments I got from the others. Due to the persistent disappointment of my past solicitors I would handle things differently this time. I would shop around the different firms before being arrested and get someone who knew the score and was willing to defend me and able to resist the telepathy of the BBC along with the Evening Mail 'foot-soldiers' as they liked to call themselves, who followed me everywhere. In May 2001, I dug out my Yellow Pages and began phoning round the solicitors of Birmingham. This was ungratifying to say the least: After the fourth or fifth firm turned down my offer to make their name in this landmark case, I decided to try a different approach. I wrote a statement claiming I had already committed an armed robbery and hadn't yet been caught, but that I expected a knock on the door any day now. I then donned my best suit and walked into town. My appearance drew gazes from all around, but this was to be expected. I knew I was looking a million dollars with my Christian Dior overcoat on to accent uate my appearance. My first port of call was Glaisyers office, which is on printing house street, off Steelhouse Lane. When the receptionist recovered from my stunning visage, I told her I needed a solicitor to represent me in the Crown Court. I was taken to see Joe Figg. Initially I met a bit of resistance from this man. First he tried to palm me off, telling me I might be better trying someone like the human rights group Amnesty International. I got a bit tetchy when he said this, but he explained that other people in Glaisyers could side with the BBC, as they cover a

THE DARK SIDE OF THE BBC. A DISTURBING TRUE STORY.

lot of the high profile cases that the firm handles. I got the distinct impression he was trying to get rid of me but eventually I won him over and he agreed to represent me in court against the BBC et al when I joked with him that I too would soon be on television. However, he was unwilling to take my potential armed robbery case, warning that such a serious charge could backfire against me. This was good. Progress was being made. I nipped into a pub down the street from Glaisyers for a spot of lubrication. As I was enjoying my alcoholic beverage the manageress was in conversation with someone and she exclaimed, "Rob me!" Which she purposely gave me the idea...robbing an individual was a much better idea than taking on a big building society or a store. I began to formulate a new plan. As I walked home people were continually being 'told' to cough all around me by the bad guys, attempting to harass me so much that I would give up the idea of my newly formed plan. I could have just robbed any of the muppets there and then but none of them took my fancy particularly as it was still early days yet and I was going to get it right this time. When I got back to my flat I sat down to consider what to do. It occurred to me that a good middle ground between robbing an individual and a bank was to go for a small place in town, far away from the clutches of Stechford Police Station and on another Station's beat. The next day I went on a reconnaissance mission around Birmingham City Centre and I eventually stumbled upon a pawnbroker off New Street. It was quiet inside but the counter was glassed off, presumably to deter those more wary felons, those who didn't have truth and justice on their side. I stood and watched this place for ten minutes, observing the girls behind the counter and the customers before heading off to contemplate this course of action. I returned two or three times over the next couple of days to keep tabs,

THE DARK SIDE OF THE BBC. A DISTURBING TRUE STORY.

make sure it felt right. But then I got some bad news which threw my plans off-kilter: My nephew Shaun was released from Winson Green Prison and where did he come straight back to? My living room, that's where. I came home one afternoon to find him there drunk as a skunk on a bottle of jack Daniels. My whole living room was awash with his clothes and things that looked as if he had deliberately thrown them all over the place in order to lower the tone of my beautiful new home and to undermine my authority as owner and keeper of such a very tasty residence. I didn't say too much about it to him because he had just come out of Prison and was entitled to let his hair down on his first day back in the real world. I just told him to make sure he tidies up after him when he sobers up the next day. But this development came at the worst possible time, and caused some serious friction to my daily routine. The worst part was that I couldn't train on the punch-bag in the morning because Shaun and his girlfriend Kerry were sleeping in the living room through most of the daylight hours. I spent three or four weeks with Shaun and Kerry living in my flat, with my plans on hold, when one day Shaun's girlfriend Kerry brought round one of her mates, Cathy Hooper. It turned out, much to our mutual surprise that I had known Cathy a long time ago and we'd been quite good friends, we'd even gone to Glastonbury rock festival together once as part of a group. I was surprised to see her again, especially turning up at my flat with no warning, so we sat down, started chatting about the old times and had a few drinks. I told her what was going on in my life, she told me her news, including that she was now on brown. Oh dear! We saw each other frequently for a few weeks, I even decorated her house for her at one stage, after which we spent most of the time drinking together at her new styled home and playing with her little son, Josh. He cost

THE DARK SIDE OF THE BBC. A DISTURBING TRUE STORY.

me a fortune in money for toys and sweets but he was a good little kid so I didn't mind at all. And then when she went to her mum's to do her turkey, in order to get off the heroin, I stopped at her place to keep my watchful eye on it and to get rid of the other heroin addicts that occasionally frequented the place. Also, because Shaun and Kerry were at my place, I couldn't focus myself on my mission to get arrested, so I stopped at Cathy's to give myself some space and to focus on doing just that as well. I stayed there for nearly two weeks, didn't see anyone, didn't touch a drop of the holy falling down water, just spent my time thinking and planning and focusing on the task in hand. The good voices helped me along, kept me on track by saying such things about me that I was "Mad" or "crazy" whenever I lost my focus. Eventually I set the date for the robbery as the next Wednesday coming. I cashed my benefit book on Tuesdays, so this seemed a good time. As I was ambling through town thinking about what the future would hold for me, I bumped into my mate Peter who told me he was in court himself next Wednesday and he was probably going to be sent down. I agreed to go along to offer moral support, I'd have a pint after, then I'd do the robbery. Of course things turned out different to that as always. Pete got bound over to keep the peace and in the mood for celebration we went out on the town. I should have known I'd be in for a telepathic kicking from the BBC and Evening Mail for not going through with my plan, and they certainly had it in for me that night. All the old tricks came out, and the people around were all taking their opportunity to abuse me, I knew it was coming as I even told Peter and I deserved it I suppose but I wasn't fazed. I was out for a good time, and that was what we were having. Until we ran out money and I realised my bank cards were all back at Cathy's. We got a taxi to her place without even having the fare to

THE DARK SIDE OF THE BBC. A DISTURBING TRUE STORY.

pay for it and I went into the shed to find the keys which I had hidden in a boot there for Cathy when she returned some day. They'd gone. Cathy must have come back, picked them up thinking I'd gone home as we'd originally planned, and then gone off somewhere herself. Me and Pete were standing in her back yard feeling a bit sorry for ourselves thinking it was all a conspiracy to ruin our day when a geezer also called Tony arrived. We chatted for a bit and it turned out he was another old mate of Cathy's, although he hadn't seen her for months. It also transpired that he'd come round earlier that day with some bread and ciggies he'd found in his possession, but she wasn't there then either. It was just her dad in then, and he'd passed them on to him. We carried on talking for a while then me and Pete came back to mine where we phoned my dad and managed to borrow some money off him. So we headed back into town, and this time managed to enjoy the rest of the evening as we were both famous now. I got back from my session with Pete to find my flat had been raided, my headboard on my bed broken, and Shaun telling me that the police had raided the place while I was out. When I woke up the next morning Shaun repeated to me what had happened and I went over to Cathy's - she told me that apparently her neighbours had phoned the cops when they saw the other Tony there, knowing that he was involved in some dodgy business and probably passing over some stolen goods. When the cops got hold of Cathy and started asking questions about this Tony bloke she jumped to the obvious but completely wrong conclusion that they were after me. As it clicked with me what had happened, I began to explain but quickly realised - this was a potential court case! I apologised for trying to lie to her and assured her that it actually had been me. I ran back to my flat where a WPC had left a calling card with the police station's

THE DARK SIDE OF THE BBC. A DISTURBING TRUE STORY.

number on it. I gleefully phoned the number and arranged an interview at Bromford Lane where I would hand myself in. This would take place in several weeks time. My court case thus assured without me having to even break the law myself, I was feeling good. The only missing part was the book. I had started writing this myself so many times and it had always fallen through. This time I would get someone else to do it for me. I put an advert in Small Heath Job centre on 10th August 2001 and straight away the responses flooded in. Most were from other parts of the country though, which seemed like it might cause problems. There was only one from Birmingham, a freaky hippy student kid called Michael Miller. I met him with Pete in the Burlington Hotel on New Street in town after he phoned up, and after looking at my options I decided he was the best bet. Another Brummy, after all, would know all about the BBC's activities. We arranged to start writing my story on a Tuesday evening, but when I got to the address he'd given me to pick him up in my car from, there was no sign of him. Obviously, I thought, the conspirators had got to him and he had given me a dummy address. Disheartened, I started writing the book myself for nearly two weeks but on a whim I phoned Michael again just to see what his excuse was. It turned out that I'd gone to the wrong address, thinking it was the block of flats next to his house. Apparently the conspirators had tried to dig their claws into his mind though. He himself had heard telepathic voices telling him to forget about the idea of writing the book, that I'd written him off. He experienced this 'telepathy' two minutes before the time I was supposed to pick him up from his house, when I myself was in the immediate vicinity. I asked him straight out if he had been got to by the BBC and their counterparts when I phoned him and he told me the truth right away, that they had tried to control

THE DARK SIDE OF THE BBC. A DISTURBING TRUE STORY.

him using 'telepathy'. He also told me that he thought the use of 'telepathy' in this way was not a good idea and then, because of that, we decided to write the book together. We met up and did a couple of evenings' work on the book before my interview came round for the stolen goods. I was picked up from my flat by a WPC and on the way I asked her whether giving a "no comment" interview would mean I would definitely be charged. I was told it would. I then decided to give a "no comment" interview. However after the interview the desk sergeant told me the charges were being dropped due to lack of evidence. Cathy was going as a witness against me, so how could there be a lack of evidence. He must have known what I was up to from the start or else he would have had me charged. I left the Police Station somewhat disheartened and made my way into town where I had arranged to meet Des in the 'Australian bar' on Broad street, otherwise known as 'Walkabout'. I told him the sad story of how the Police had refused to charge me, before drowning my sorrows in a few pints of the 'holy falling down water'. A few days later I was in Small Heath and still feeling sad when I slumped down into a chair in Lisa's Diner down the road from my flat. I indulged my self-pity for a moment over a jumbo breakfast and coffee before spotting Cathy walking past with a workmate of hers called Brian. They came in and sat down with me, and she proceeded to accuse me of robbing and cashing her benefit book then putting it back in her house as if nothing had happened. I guess she was basing this on my previous 'confession' about the stolen goods but it was still a bit stinging to be accused of stealing from one of my mates. I assured her of my innocence once again before realising that this could, once again, work in my favour. I told her that I did not cash her benefit book at all but that I needed to get arrested to expose the BBC , so I persuaded her to tell the

THE DARK SIDE OF THE BBC. A DISTURBING TRUE STORY.

Police that I did do it. We went to the Post Office and verified that the missing cheque had indeed been cashed, then we came to my flat where she phoned the police and told them what I'd supposedly done. But yet again the efficiency of our law enforcement services shocked and disappointed me: There was no CCTV of me at the Post Office, so they eventually decided not to press charges.

      This constant failure for the pettiest of reasons was getting to me now. Obviously I was going to have to commit some crimes myself, and make sure there was no doubt that I did it. The constant pressure of the 'voices' in my head compelling me to get myself arrested at all costs now was making me act under duress anyway. I bought another car and went on a spree, robbing garages of petrol and driving off. Their video cameras would provide incontrovertible evidence of my guilt, plus my car itself was a whole caseload of offences such as no tax, no insurance and no m.o.t etc. So I spent a while driving around Birmingham with my free petrol and no insurance, tax or MOT, but at the same time me and Pete began interviewing other receivers of telepathy. We managed to get three interviews done before Josh, Cathy's five year old son somehow knocked our video camera off its tripod and damaged both the camera and the tape. In September my car got written off by one of my friends who shall remain nameless ....PETER FLYNN!

This was a wounder though because it meant my weeks of unlawful activity were for nothing as I wouldn't now be captured driving the 'hot' car. A few days later, again acting under duress, I bought a second-hand Escort to replace it and got it registered in my name so that it was traceable to me in the event of any crimes, but this time I didn't do a robbery, as I got pulled over on 9/10/01 for simply not having any tax on the car itself, which was all part of the plan. I was given a

THE DARK SIDE OF THE BBC. A DISTURBING TRUE STORY.

producer, i.e. produce your documents in a week or else. The seven days passed and proceeds to summons me began on 16th October. On 26th October I wrote a letter to my would-be solicitor Joe figg explaining to him in writing of my attempts to get myself arrested in order to expose the BBC and the Evening Mail for their illegal activities. He was kind enough to write back in acknowledgement of receipt of my letter. I also decided to write a letter of complaint against the BBC and the Police, to the Chief Constable of the West Midlands Police, Sir Edward Crew, so I fired off a letter to him dated the 10th August 2001. He then passed my letter on to the Chief Superintendant who in turn appointed an investigating officer and sent me a reply to my letter. The reply he wrote to me was a few words to acknowledge receipt of my letter and to inform me that an investigation was underway. I was told that I would be contacted about my complaint in due course. Not so. Instead of investigating my complaint as promised, I was not even approached by the so-called investigating officer to make a statement about my troubles and strife; instead, he or she took it upon themselves to make phone calls around the medical profession to tell them that I had been writing letters to the Police.

In short, he was trying to get me covered up by the medical profession again, which was totally unacceptable. I found this out by my CPN (community Psychiatric Nurse) Mick O'Hanlon, who told me he had received messages to that effect. I had to think fast so I just told him that I was complaining about something my nephew Shaun had done, which was bull-shit, but it worked. Mick just said to me "oh, it's nothing to do with me then?" To which I replied "no!"

THE DARK SIDE OF THE BBC. A DISTURBING TRUE STORY.

I got out of that little jam but in order to stay out of it, I had to write another letter to the Chief Superintendant at Police headquarters in Lloyd house, Birmingham City Centre to tell him that I wanted to drop the charges as the matter had been resolved to my satisfaction. I had to do this in order to get the Police off my back and wrote another letter to my would-be solicitor Joe Figg and told him all about it. He did not reply this time himself as he was on sick leave in hospital.

It made me wonder if he was being terrorised by 'voices' in his head, which would not have been unusual. Robert Maxwell was still at large somewhere but by now there were others taking his place for him like some kind of team offensive.

MORE ARRESTS, MORE TRIALS.

November 7th turned out to be a nice day actually. I flipped out of bed in the usual manner and tuned in to 100.7 Heart f.m, (the biggest radio station in the midlands). I couldn't be bothered to make breakfast that morning (even though

THE DARK SIDE OF THE BBC. A DISTURBING TRUE STORY.

it could have been only Weetabix). I just fancied eating out basically. So after a good wash and shave I decided to head down to the local Asda supermarket on coventry road for a slap up breakfast. I knew by the reactions and comments of people around me that I was thought and sight broadcasting but it didn't deter me in my course of action. I was a hungry man. It was quite full of people in the restaurant at that time in the morning, (about 11 o'clock), so I was lucky to find an empty table to myself. The breakfast was o.k but all the snide comments around me were starting to piss me off. I retaliated by reacting a bit moody towards them, knowing that they wouldn't even do anything about it. I began to enjoy it again a bit in a strange sort of way, I was different from every body else, like I was on a special mission. I didn't like the situation that everybody was buzzing off me, and I was in a bad mood really. I finished breakfast and proceeded to walk around the store to see if I could find any bargains. Then I noticed that a lot of people were strategically 'appearing' around me and they all looked menacingly at me, and they were all talking into mobile phones. Obviously seeing some kind of nasty conspiracy taking place, I decided to confront them as best as I could. They were making themselves blatantly obvious as they followed me around the store, glaring at me constantly. Then I realised who they were. It was the Evening Mail gutter press and they had involved the Asda security staff in their conspiracy. There was about seven of them altogether, mostly females, not many men, and they were all trying to intimidate me into not committing an offence that would eventually expose their illegal activities in court. Some joke! Most of them were ageing old women that would have been better suited being at home doing some knitting and some of the men bore resemblance to that of a stick insect. I had to laugh

THE DARK SIDE OF THE BBC. A DISTURBING TRUE STORY.

at them in their persistence of my demise until they made me want to confront them, to see what they could do. I decided to let them know telepathically, that I was going to shoplift. This made them react more furiously in their malevolence, which amused me to such an extent that I decided that I really, really would do it. I would call their bluff.

At this moment in time, I would get myself arrested for shoplifting again and use the court case against them all; only this time I knew what I was doing, as well as the procedure to do it. I was fearless to the point of total confidence. I would commit the act of theft, but not to steal from Asda or permanently deprive them of anything, I would do it purely to get myself arrested. And see what would be done about it! The good guys in my head urged me onwards. I selected two bottles of whiskey because they were more expensive than most of the other stuff and would make the security staff more obliged to capture me. I was daring them to have me arrested, taunting them even. I put the two bottles of whiskey into my jacket pockets right in front of them more or less and proceeded to walk out of the building. They were still talking into mobile phones when they approached me in the entrance foyer, but the leader of the gang who stopped me had a smile of resignation on his face. He knew I could not be frightened into submission. He asked me if I had concealed two bottles of whiskey about my person and I beamed back at him as I replied "yes, you know I have". These smiles on our faces would have been caught on the video cameras and would have shown proof of some kind of unusual conspiracy but somehow later on, after the police told me there was video evidence against me, I was told that there was no video evidence and that the said tapes had been recorded over since. Bullshit! Anyway I was taken to the police station by

THE DARK SIDE OF THE BBC. A DISTURBING TRUE STORY.

the arresting officer, PC 8570 WYKES, made a no comment statement (I knew by now that if I mentioned a conspiracy involving the BBC or anyone else who was involved, I would surely end up swept under the carpet by Police and 'covered up' inside a mental hospital), then I was charged and bailed to appear before Birmingham Magistrates Court on 9/11/01 at 09.45 hours. Things were looking good again!

DRINK-DRIVE NO 1.

THE DARK SIDE OF THE BBC. A DISTURBING TRUE STORY.

On the 13th of November I had to pick Michael Miller up from his house in Hockley. It was Tuesday, which is payday for me, so I bought some drinks for us to have while we were working and got to his place at 7:30. We nipped into the Australian bar on Broad Street for a quick bevvy before heading back to my flat to start work. I realised while we were in the Australian bar that if I had more to drink at mine I'd definitely be over the limit and told Michael I'd give him the cab fare home afterwards. We had a lot of work to do: It was a crucial time with a court case pending, so we needed to get the book up and running in order to get it up to date. Michael was doing most of the writing, working from the notes he'd taken, which I would then add to and edit myself. I was drinking cider and lager while we were working. Although I wasn't feeling particularly pissed I knew I was technically over the limit and reminded myself that I was going to get Michael a taxi home. But no sooner had I started thinking this, than the voices in my head began to pressurise me into driving the car illegally, as they always treated any respect I showed for the law as a weakness. The voices in my head were intolerable at times and I knew I would succumb to them eventually. But when they started giving me electric shocks, causing spasms throughout my body, I quickly gave way to their demands. In a state of what could only be described as duress, I agreed to drive Michael home despite his own protests. He seemed concerned about my decision but I think he could tell I wasn't going to listen to him: I was compelled. We drove back through the city centre and I dropped Michael off, arranging to meet the next week as usual. I then headed back home via the same route and had just passed Digbeth police station when some drunken geezer leapt out in front of me, nearly killing himself. As I was swearing at him after stopping the car, he

THE DARK SIDE OF THE BBC. A DISTURBING TRUE STORY.

jumped into the passenger seat and asked me to drive him home. Fair enough I thought, after extracting a fiver from him. I did a u-turn and headed towards Hagley Road where he asked me to go. After a couple of minutes driving, some guy pulled his car up next to me and started motioning for me to pull over. I shooed him away thinking he was harassing me and carried on driving, but he was persistent and eventually I noticed the badge he was wearing. He was a copper in an unmarked car. I pulled to the curb and he gave me a breath test which obviously I failed. At this point the drunken guy I picked up had jumped into my seat and was busily trying to start the car himself. The copper sent him on his way before confiscating my keys and taking me down to Steelhouse Lane Police station. To me the whole situation seemed a bit dodgy. I wouldn't go as far as to say it was definitely a set-up, but the combination of events was a damn-coincidence. Anyway, I failed the computerised breath-test and was refused bail temporarily. I was only let out at six the next morning, but the Police kept my keys and would not return them to me until I was within the drink drive limit again. I wasn't too dismayed anyway at these events as I realised the voices in my head had got me another court case with which to expose their illegal activities once again.

THE DARK SIDE OF THE BBC. A DISTURBING TRUE STORY.

DRINK-DRIVE NO. 2.

THE DARK SIDE OF THE BBC. A DISTURBING TRUE STORY.

Within a relatively short time of getting myself arrested for drink driving, I managed to get myself arrested again for the same offence. I don't remember a whole lot about that day except for the ending really, simply because (as I realised later), the BBC could black me in and out of consciousness as they pleased, even if I was perfectly sober and alert. (The full details of this phenomenon are described in Drink Drive no. 4). I was out with Des again and we had a few tots in Small Heath before I drove us into town and up to Broad st, where all the action is. We were both chilled and enjoying ourselves as usual, but after about four pints, I was 'blacked out' by the flick of a switch and from there onwards the BBC were in control. After a long while of being unconscious on my feet, I suddenly came back to consciousness in the early hours of the morning and I was standing in the road next to my car (which I had been driving apparently). From sudden darkness to normal vision, for a split second at the touch of a button, all I saw was Des on the other side of the street to me looking well apprehensive. Then I was 'blacked out' again before returning to consciousness again a few seconds later to find myself standing before three serious looking Policemen. The next thing I knew was that I was in the Police station being charged for Drink Driving again. One of the WPC's there told me that I was unconscious on my feet and I soon found out that I had been punched in the mouth as well that night by some real brave soul who Des told me later could have been a doorman. This was the first time that the BBC really started to take over my mind and put me in a state of blind unconsciousness even though my faculties were functioning well beforehand. They could do this to me even when I hadn't taken any alcohol, so I was never safe from their mind control. I was bailed from the Police station in the early

THE DARK SIDE OF THE BBC. A DISTURBING TRUE STORY.

hours of the morning when I collected my car from near Broad St, and went home to bed.

IN LONDON FOR THE NEW YEAR, 2002.

I spent Christmas 2001 with my Mom. She didn't approve of me drinking too highly, so I spent a few days with her attacking the turkey she had roasted and generally pigging out. I saw the most of my immediate family over that period and when Christmas had gone, I went back to my flat alone. Shaun and Kerry had long gone by now, they had got their own flat together in Small Heath, so I had redecorated my own home again, this time in spice red for the living room. I had already bought a brand new computer to write and print my book on, a new workstation and office chair along with an office desk, so the place was looking well up-market. I didn't have a lot of cash left after the Christmas period but I had my trusty ford escort parked outside, so I decided on a little trip back to London again to see in the new year. I couldn't get hold of Pete so I called round for Des on the off chance that he would be in. He was. I asked

THE DARK SIDE OF THE BBC. A DISTURBING TRUE STORY.

him if he fancied a couple of days down in London for the New year and he was up for it so we set off within a couple of hours. We stopped off at Asda in Small Heath to obtain some liquid refreshments for the journey down, where we bumped into Pete Edmead and another pal of ours, Mickey O' Dowd, talked for a short while together before we headed off on another excellent adventure. I didn't care about the BBC, the Evening Mail or Robert Maxwell who was hiding somewhere behind the BBC, I just wanted a few days holiday from the glare of 'publicity' and some time to party-on in the Capital for the new year. Unfortunately, the Birmingham team of broadcasters did not like the idea of me going down to fraternize with the 'other side' and so they left me thought and sight broadcasting from the moment we began our holiday, to the time we arrived back in Birmingham. I mean constantly. It was probably the work of Robert Maxwell and his team who were out to ruin my enjoyment, but I didn't even think about it enough for it to worry me. I didn't care anyway. I was too strong-minded to be adversely affected. We made a few more stops on the way before we left Birmingham and I could tell what was happening by the look on people's faces when we stopped and by the comments they made as they came into my vision. I even asked Des to confirm my suspicions by asking "how do people know when to make comments about me, is it because they can see what I see?" He smiled and replied "yes it is". I drove on regardless.

We arrived in London in the early evening and straight away tracked down Patrick's new flat (he had moved since my last visit with Lynn Vale some years before), which was now in Shepherds Bush. He wasn't in when we got there but I only wanted to leave my car outside his house for the time being, so that me and Des could go out drinking in the West End. I did just that

THE DARK SIDE OF THE BBC. A DISTURBING TRUE STORY.

before starting off our pub crawl in the South African Bar in Shepherds Bush where all the men seemed to be smiling knowingly at me and the women were up dancing around me as they probably liked to see themselves 'on camera'. I didn't get too involved myself because I knew by now that it was becoming hard for me to get off with a girl who knew I was telepathic, and when I did get off with a girl, she would receive instructions from the studio shit-heads to give me a dummy phone number or something. I had to be very selective. As it happened, a good time was had by all, with me and Des sleeping on the floor of Patrick's living room at the end of each night. We saw the new year in as planned but the only hiccup so far was when the Police were called out to us by the manager of a restaurant we were in after I had called him Osama Bin Laden. He deserved it for the cheek he gave us. The Police didn't do much as we both decided to take our good custom elsewhere anyway. It was on the morning of our intended departure back to Brum that something untoward happened to me.

We had just got up with the early morning sun and were planning our route back when I was suddenly struck down with a searing pain in my right leg. It lasted a couple of minutes but it was very intense. I stretched and massaged it in that time but to no avail. Des said it was probably gout but it made me wonder if it wasn't electrically orientated. As it was, we left Patrick in London after I told him I was writing a book and headed off towards Birmingham, stopping at every service station on the way back. I was exhausted really after our stint in London and was too tired to face the crowds we encountered in the service stations nearing home so I remained in the car towards the end of our journey, while Des went in to top up our supply of pop or water. I was so

THE DARK SIDE OF THE BBC. A DISTURBING TRUE STORY.

exhausted that I don't think I went over fifty miles per hour on the way back, whereas on the way down, I was hitting around ninety. We arrived back in Birmingham at about midday and were greeted by an Evening Mail man as we drove through Sheldon, just before Small Heath, who stood out on the road and held a small camera up to us with his face looking well agitated, just to let me know that they were back on the case. Or at least the 'bad guys' were. I just laughed at him as I pointed him out to Des. Soon enough, I dropped Des off at his flat and told him I was going home to bed for a few hours before driving up to the end of his road, a cul-de-sac and reversing back round a bend in order to come back on myself. Just as I did this, I was again stricken with the same searing pain in my right leg as had happened in London. It was so bad that I had to slam the hand-brake on and hobble out of the car, leaving it parked in the middle of the road. Stretching and massaging proved useless as I hopped around on one foot, my distress being ogled at by a couple of young kids. This was Maxwell and his team's way of punishing me for returning to Birmingham from the 'other side's' manor, London town. They had been trying to drive me out of Birmingham for a long time now, especially because I knew that it was Maxwell that was involved in the dark side of television. An Asian man and woman came out of their house and studied me for a few seconds before the woman shouted out in an alarmed voice "and he stays in Birmingham". This was her way of letting the bad broadcasters and the good people of Birmingham know that I was not going to be driven out of town no matter what. I was ruling the rabble once again. This was my task.

THE DARK SIDE OF THE BBC. A DISTURBING TRUE STORY.

THE DARK SIDE OF THE BBC. A DISTURBING TRUE STORY.

THE FIRST COURT-CASE 2002.

I had been in and out of the Courts on many occasions awaiting trial for the alleged offences, but the time finally came round for the first trial to begin. It was my big day. At last I would expose the illegal activities of the BBC and finally get my life back to myself. I had sacked Joe Figg from Glaisyers Solicitors because he had betrayed me with his unwillingness to represent me at a trial in court and had decided to represent myself this time for the theft of the whiskey allegation. I was capable enough being an ex Grammar School boy but I was also in the process of employing the services of Brian Pugh from Carvers Solicitors to represent me in the near future who was concerned enough at the time to tell me that I was "up against the establishment". For my trial, I had prepared exactly what I was going to say in a statement to the court, I had gathered four witness statements which I was going to show to the court, naming other people who had witnessed the existence and misuse of telepathy and I was raring to go. The trial was to be held in the Magistrates Court and I duly arrived early on my own and waited outside the court for me to be called in so that the proceedings could begin. After a short while, Des turned up out of the blue (he was supposed to be at work), and we sat patiently outside together full of expectancy. There were no newspaper men or women (even though I had written to all the nationals) and there were no TV cameras outside, or even the local radio station I invited along, Heart FM. No-one was there at all. I was not too bothered, but I became slightly suspicious. The time

THE DARK SIDE OF THE BBC. A DISTURBING TRUE STORY.

came for my trial to start and I was called into the court room by the court usher and stood before the magistrate, who was a woman of about 50 years old. The prosecution made their case against me first, and then they called two witnesses against me. The first was a white female store detective from Asda and the second was an Asian male who was a security guard on duty at the time of the alleged offence.

The female stated her response to the questions asked of her, then the Clerk to the Court asked me if I wanted to cross examine her as I was conducting my own defence.

I had no intention of cross examining at first, but then I realized that I could turn the prosecution witnesses around in my favour by asking them if they had experienced or were controlled by telepathy. They were bound to say they had because they had. I asked the female first if she had ever experienced telepathy, especially at the time of the alleged offence but alas she smiled cunningly at me and replied "no!" In short, she lied under oath. The security guard was next and when I asked him, he faltered a little at first so I reminded him that he was under oath before asking him again. This time he was still a little indecisive but replied that he had experienced telepathy, when he put a mobile phone to his ear. He was nervously trying to mix up fact with fact. I should have steamed into him there and then with more questions but the Clerk to the Court intervened with some bullshit and my man was off the hook. The clerk took advantage of my inexperience. Now it was the turn of the defence, my turn.

I gladly took the stand and took the oath. My defence was that although I had indeed taken two bottles of whisky from Asda, my intention was not to steal

THE DARK SIDE OF THE BBC. A DISTURBING TRUE STORY.

from them or permanently deprive, but to get myself arrested, therefore I was not guilty of theft. This was a true legal fact which I had used before in the crown court when I was found not guilty of robbing the nationwide building society. When I told the Magistrate that I had prepared a written statement for the Court, the Clerk intervened again and told me that I could not use my written statement, nor could I produce my four witness statements, nor could I produce an extract from my book that detailed the alleged offence. In short, it was all another cover up. I was not being allowed to expose the illegal activities of the BBC and co. The reason was that I had not informed the prosecution of my intentions at a previous Court hearing, which was called a pre-trial review. I was never told that I had to. It was all part of the growing plot. I looked at Des in exasperation and he seemed a little concerned too, though he could not hear the proceedings as well as I could. I even told the Court that I had exposed the BBC over the internet some weeks before, along with the Evening Mail and none of them had done anything about it. I had employed a firm called CT Graphics to design a Web Site for me for the expose, but the prosecution said that the BBC and co probably didn't know that the web-site was there.

Some excuse, but I thought next time I do it over the internet, I will send out letters to the BBC and co informing them of what I had done (and kept copies) so that there can be no mistake.

The farce continued on until the Magistrate decided to retire to her chambers to decide if I was guilty or not guilty. Des and I went outside the Courtroom to await my fate. After about fifteen minutes, the Clerk came out and told me that the Magistrate was seeking his advice in her chamber, along with the

THE DARK SIDE OF THE BBC. A DISTURBING TRUE STORY.

prosecution but he could not tell her anything in secret and would inform me of everything he told her when Court resumed. Bullshit!

When Court resumed some fifteen minutes after that, I was told nothing of what went on in chambers as I stood before the Magistrate, who in turn mumbled out to me that because I had admitted the actual 'act' of theft, I was guilty as charged but she was postponing sentencing in order for a medical report to be done, (after which I would be packed off to a mental hospital again). Another cover up had been instigated.

THE DARK SIDE OF THE BBC. A DISTURBING TRUE STORY.

DRINK-DRIVE NO 3.

I didn't have a fat lot of money so I decided to have a couple if pints in a local pub close to my flat called the Sampson and Lion, on the off chance that Peter Flynn would be there as he does frequent that pub more regularly than I do. As it happened, Pete was already in there when I arrived so we sat together at a table near the pool table as usual, and partook in a few beers and a few games of pool. The evening passed by without too much incident really, even though I slowly became aware of the fact that I was thought and sight broadcasting (this was becoming the irritating norm now every-time I started to have a few pints),

THE DARK SIDE OF THE BBC. A DISTURBING TRUE STORY.

but it was because I was still buzzing away so confidently and ruling the roost so to speak that no-one hardly commented into the micro-phones. It did seem more or less like a normal evening in a pub for me, until that is, a face from the past entered the pub, which hallowed the arrival of another would-be conspiracy. It was one of the psychiatric nurses present at my unlawful detention in Highcroft Hospital some three years ago. His name was Idris, a short Welsh guy of about thirty something. He was the nurse that restrained me on the floor the one time I got into a fight when I was locked up on Meadowcroft ward in Highcroft Hospital. This was probably one of the high points of his career as he seemed to relish in the knowledge, even though I hadn't put up any sort of resistance. I recognized him straight away as the one who was present at the cover up in the Crown Courts, when I had gotten myself arrested for attempted robbery at the Nationwide Building Society in Small Heath. I told Pete that I knew him and Pete asked who he was. I just took the piss really by replying that he was just another corrupted psychiatric nurse. I didn't really care if I was broadcasting or not, though I guessed I was, (which is the reason why I said it in jest), but from the paranoid reaction I immediately witnessed from him (as did Pete), I then knew I definitely was broadcasting. Idris made his way to the bar looking unconvincingly confident, like a man on a mission, then disappeared into the crowded bar.

I continued conversing with Pete for a short while before Idris appeared at our table with a pint in hand for me as he sat down to join us. His peace offering was gladly accepted, though it masked the true intent of his appearance. Idris had a bit of a bad reputation at the Hospital for his excessive drinking habit and tonight I witnessed this excessiveness for myself. The trouble was, he was out

THE DARK SIDE OF THE BBC. A DISTURBING TRUE STORY.

to get me drunk as well. I was drinking moderately, as was Pete, but Idris relished in the act of mixing his drinks in order to get drunk. He was contagious. He seemed to be enjoying himself to such an extent that after continually trying to force his drinks upon me, I relented and joined in the game. I was only sipping all night until then so it was quite late in the evening when I actually relented. Towards the close of the night, Idris tried to force the issue of going to the Gary Owen Nightclub in Small Heath. The reason for this was because I wasn't drunk enough for his liking. But I was adamant that I would stay in control. By last orders I was quite willing to go to the Gary Owen with Idris and Pete and even tried to cajole Pete into lending me the money to go with them. Pete was not amused at all by the sudden intervention of Idris and did not seem too happy about lending me the money to go, or even about going with us himself. He was aware of something fishy going on and even disclosed to me in private that he was 'concerned' about the sudden arrival of Idris. I didn't take too much notice of him as I knew Idris quite well although Pete was quite right at the time. So with Idris now trying outside to shepherd me down to the Gary Owen and with me trying to drag Pete along, the three of us left the Samson and Lion pub and proceeded to my car which was parked on the opposite side of the road. No sooner had we reached the car than another incident took place that fuelled Pete's suspicions. A police car pulled up some yards behind us on the opposite side of the road and its occupants sat there as if waiting for me to drive my car. Although I was pretty sober, I was obviously over the limit and I was still thought and sight broadcasting to all asunder, but the Police were there as a deterrent for me not to drink and drive as they were watching me closely now that I had court cases pending. I was quite aware of what was

THE DARK SIDE OF THE BBC. A DISTURBING TRUE STORY.

going on but was too headstrong to be deterred by their threatening presence. I got my keys out and told the other two to get in the car as I was determined to call their bluff, but after some serious deliberation with Pete and Idris, Idris decided to leave it and headed off home.

Eventually, Pete persuaded me not to drive the car home in front of the police as I was surely tempted to do, but instead he invited me back to his house which was just around the corner for a few glasses of wine. I left the car where it was and headed off towards Pete's house, passing by the Police car on the way. As I was not now committing any crime, I decided to take the piss with the Police as I passed them. I leaned into the open passenger window and asked the male officer there why they had decided to park there all of a sudden and jokingly enquired "is this a conspiracy or what?" He made some reply before we both laughed together and I went off with Pete. We actually polished off two bottles of wine that night at Pete's house before I left at about 2am and made my way home.

Sunday soon came around and I was up pretty early as usual knowing that I had to give myself enough time to sort my affairs out indoors and still have enough time to meet Des at his flat at 12 o'clock midday. We were going to have a look at an exhibition of some properties for sale on the Spanish Costa Brava. Apparently, these properties were being sold direct from the builders in Spain so there was to be huge savings to be made on many purchases. Des was interested in buying some land and building his own house over there whereas I, if all went to plan, was more interested in buying an actual property eventually.

THE DARK SIDE OF THE BBC. A DISTURBING TRUE STORY.

Of course I was thinking more in the long term than Des, but the idea of owning a Spanish property was definitely appealing to me so I was happy to go along to investigate as there was no obligation to buy involved in any case. My whole life was still being kept on hold by the BBC at this time, so long term forward planning was all I had to look forward to anyway. I was more confident of making the rest of my life a success after all the shit I'd been through so the end result of fame (which I already had) and fortune was not unreal to me. I arrived at Des's flat about fifteen minutes early where we discussed his recent big win on a football bet he had done at Ladbrokes over coffee. He estimated his winnings to be somewhere in the region of £170 and as the Small Heath branch was open on Sundays, we decided to collect his winnings on the way to the Exhibition, which was at an upmarket hotel in Merry Hill, Dudley, just outside Birmingham. We duly arrived to collect the big winnings soon enough and to both our delight, the estimate was near enough correct. Des was handed over the tidy sum and we left the building in renewed high spirits.

The drive to Merry Hill was an adventure in itself even though we both had a fair enough idea of which way we were going. It wasn't until we were actually in Merry Hill and I got to thinking about the return journey back to Birmingham that Maxwell confirmed my suspicions about the origin of the excruciating pain that I experienced in my legs that time when I had decided to return to Birmingham from London after our three days of revelling to welcome in the new year. I had just approached a large roundabout towards the end of our journey and saw a sign post indicating the way back to Birmingham which made me think about the return journey. I was now travelling at about 30-40m.p.h when his attack began and continued on for about 5-6 minutes. It was

THE DARK SIDE OF THE BBC. A DISTURBING TRUE STORY.

my upper right leg that bore the brunt of his nasty little attack and although the pain was serious enough to have easily caused me to lose control of my vehicle if I'd have had to break hard for any reason, I didn't even bat an eye-lid in response as I continued to make conversation with Des as if nothing was happening. Of course I was really screaming with pain in the back of my mind but I knew from experience not to show any signs of pain or weakness, so after a minute or so, I had relaxed enough to such an extent that the pain in my leg was under control enough to seem barely significant. The pain itself could have been compared to that of extreme cramp so you can imagine what I was going through at the time. This was about as brave as Maxwell ever got from the safety of his studio-type hideaway.

I couldn't really let on to Des what was really going on as it would have been pointless to get him alarmed at the situation as well when there was absolutely nothing that could be done about it. Eventually, as the pain continued on, I began to goad Maxwell to do his worst, describing him to myself as the sort of cowardly scum that he was. That was probably why the attack went on for as long as it did in comparison to the time when I returned to Birmingham from London after the new year, when I was not so sure about the origin of the pain. But I hated him so much now that I would rather show him my full felt hatred and take the fullness of his attack than show any deviation from those feelings. My animosity towards him and his cronies hiding behind the BBC won the day as usual. The pain subsided and slowly disappeared within a few seconds of our arrival at the Hotel where the exhibition was taking place. We alighted from my car and I secretly stretched and massaged my leg Just to make sure that all traces of pain were gone. We both entered the Hotel via the entrance in the

THE DARK SIDE OF THE BBC. A DISTURBING TRUE STORY.

foyer and still laughing and joking together about how neither of us looked dressed up enough to look as if we could afford to buy a Spanish property, we descended into the rooms reserved for the exhibition itself, with Des still blissfully unaware that we both could have been killed in a car accident only minutes before. We stayed and viewed several properties for half an hour or so, raiding the coffee and biscuits relentlessly just for the hell of it and pissing ourselves laughing at each other in the process. We made arrangements to go on a 4 day tour of Spanish properties on the Costa Brava at a cost of £115 each and both left with an information pack of all the properties available in that region.

    The drive back to Birmingham was a lot easier on the way back as we were both now familiar with the route. We stopped off at a Harvester pub for a pint of the holy falling down water on the way. Once inside the pub, I scanned the clientele and kept my ears working overtime for any signs of telepathy in action. There was none. We both finished our pints together and left the pub in Dudley in order to head off to the next pub we had decided to visit which was on the hagley road in Edgebaston, Birmingham. We didn't reach our next destination though as I drove past it by mistake, so Des phoned up his mate (who was also called Des and also a bricklayer) on his mobile and arranged to meet him in one of his local pubs called the Red Lion in Acocks Green. We arrived a bit early and had a pint each before Des no. 2 arrived on the scene and bought another drink each for the three of us. After that, I could tell from my observations that telepathy was now in action (it usually did start after about two pints these days) and some well rounded wannabe film star was the focus of attention.

THE DARK SIDE OF THE BBC. A DISTURBING TRUE STORY.

I just knew from her sudden acting 'spree' that she was going to make a bee line for me eventually and I watched her psyche herself up before she finally landed upon us.

We boys knew the *craic* and we all took the piss out of her as she ducked and dived around like she was the bee's knees. Unfortunately though, she latched on to me to such an extent, after giving me a big kiss on the lips (which I didn't like at all), that when we decided to leave for pastures new and classier, she worked her way into the car with us and sat on the back seat with Des. I was driving and Des no.2 sat in the passenger seat as we were going to drop him off home with the intention of meeting him later at a discotheque, which was on at the Westley Arms Hotel. In the meantime, me and Des were going to go down to one of our local pubs in Small Heath, the Samson and Lion for a couple of pints and meet up with Pete.

I allowed the wannabe film star to come just for a laugh, but slowly began to disrespect her when I detected her snide intentions. There were two other blond haired women in the pub that afternoon that I had a little fracas with the previous week but when they saw me in a bad mood, they were all over me like they preferred me that way. I lost interest in the three of them eventually and they all disappeared soon enough from the pub, much to Des and Peter's relief that the three femme fatales had gone. Us three boys then continued our revelling throughout the afternoon and eventually met Des no.2 in the Westley Hotel as planned.

We all had an excellent time there to such an extent that at the end of the night, Des, Pete and myself decided to go on into town in search of more of the same. I was too drunk really to drive so Pete took over as driver for the night.

THE DARK SIDE OF THE BBC. A DISTURBING TRUE STORY.

He drove us into town and parked the car on Bradford street, but all the pubs we tried were already closed or just in the process of closing. Des reckoned that the best pub to have a chance of getting into was the Kerry Man in Digbeth. We went there as well but they wouldn't let us in at that time of the night. A Sunday night at that. We had failed in our mission for more fun so we decided to go home. The three of us returned to the car some minutes later and I got into the driver seat just to wind Pete up that I was going to drive home even though I was well pissed, but I was also really going to show him how to operate the lights etc as it was now quite dark. Pete was holding the keys anyway so I didn't have any chance of driving at all really, but within a couple of minutes of me getting into the car, a Police car pulled up behind us and after a couple of seconds, a copper got out and made his way towards our car. He was joined by his colleague who emerged from the passenger side.

We all panicked a little at this (as you do), but because I was not in possession of the car keys, I got out of the car to face the old bill. Straight away, the first copper demanded the keys off me saying he would let us go if I gave him the keys. Of course I didn't have any keys on me so I told him that, but after a few seconds of what was really a stand off, the same first copper said "right, that's it!", then he grabbed hold of me and, along with the second copper, they both bundled me into the back of their van. I didn't put up any resistance because I hadn't committed an offence but these chaps didn't seem to care really, they just wanted to nick me now even though they knew they were in the wrong. It dawned on me at a later date that it was because I was always thought broadcasting whenever I had a few beers that the Police were always following

THE DARK SIDE OF THE BBC. A DISTURBING TRUE STORY.

me around and so were always in the position to nick me if they so desired if they knew I was over the limit. Some-one tell me that has to be illegal!

The car was searched and the keys were found on the floor of the passenger side (obviously Pete didn't want to be found in possession either) and I was transported off to Steelhouse lane Police station, where I was duly left to my own devices in a small cell. I just slept the time away until I was roused from my slumber and taken before the desk sergeant in the early hours of the morning. I was charged with being drunk in charge of a motor vehicle etc, and refused bail for the first time. Although I didn't know it at the time, it now transpired that the Police conspiracy to follow and arrest me for drink related driving offences had reached its crescendo, and I wasn't to be given bail anymore. This was my third drink drive offence. I hadn't ever had an accident or injured anyone in any way but I was still foolish for getting involved in scenes that could always backfire on me, especially when the Police were not exactly helpful towards me. I slept again for the rest of the morning until I was awoken again at about six a.m to get ready for my court appearance that day.

It was a Monday morning and I was taken across to the holding cells underneath the Magistrates Courts. I was kept there for some time to await my appearance and to await the arrival of my solicitor, Brian Pugh from Carvers Solicitors. The time spent in the cell with three others was pretty uneventful until the time came for me to see my solicitor. I was handcuffed and taken out of my cell and off to the interview room whereupon I was met by a female representative of Carvers (I cannot recall her name), and not by Brian Pugh himself. This was not unusual in itself but as the interview proceeded, I could detect a feeling of cool anticipation from my female solicitor. This made me feel

THE DARK SIDE OF THE BBC. A DISTURBING TRUE STORY.

slightly uncomfortable. It was as if she was just waiting for me to say something wrong so that she could punish me. In actual fact, she was waiting for me to mention the BBC or the Evening Mail. This would have made it difficult for me to get bail, as the conspiracy to 'cover up' was well under way now that I had mentioned their involvement before on the other three charges, which were one shoplifting and two drink drive. This charge made four altogether, of which the BBC and their counterparts were guilty of instigating all four. On this charge however, I was saying that the basis of my defence more or less was that I was not in possession of any car keys at the time of the alleged offence so I was not in charge of the said motor vehicle, therefore I was not guilty of the alleged offence. When I had finished outlining the basis of my defence to her (without mentioning the BBC), My female solicitor beamed at me with delight and enquired "is that everything?" I was a bit confused with her sudden joy but replied "yes".

Then she promptly sprung from her seat at the table opposite me and exclaimed aloud "that's all that I wanted to hear!" (she was overjoyed that I hadn't mentioned the BBC or their counterparts). Now I had a realistic chance of bail.

When I eventually landed in the dock before the male Magistrate, Brian Pugh was there himself, ready to address the Court.

I watched the scene from the dock as Brian Pugh made his speech to the magistrate. He came across as slow and methodical in his approach, ultimately giving me the impression that he genuinely enough wanted to get me bail. This wasn't too important to me now though as I was more interested in the expose of the BBC to go ahead. I listened with great interest to the

THE DARK SIDE OF THE BBC. A DISTURBING TRUE STORY.

proceedings as Brian went on to say that in this case, I had not mentioned the existence and misuse of telepathy by the BBC so therefore I didn't appear to be delusional anymore and that the medication I was taking must be working. I felt deflated. I realised then that my hopes of a return to a happy normal life were not to be achieved with the help of Brian Pugh at the helm. I watched as the magistrate seemed to relax more now that the pressure was off him but my mind had turned elsewhere now to analyse my choice of solicitor. He was a good man, from the old school like myself but I remembered then about the time my first case for shoplifting was put off for psychiatric reports and Brian had told me later that he wanted the reports to be used on the other cases I had pending. I had told him then that that would not be a good idea as it would not help me in my cause, to which Brian had replied rather worriedly "but it will help me". I realised then that he was being pressurised and co-erced into taking part in the cover-up, either by other solicitors who had something to hide, or by the Evening Mail 'foot-soldiers', or by BBC telepathy. He had been taken over by the conspiracy against. It made me sad but I knew then that Mr. Pugh had to go. I decided not to bother with the services of any solicitor anymore, I would do the rest of the cases on my own as I had done previously but I was more educated in the pitfalls of handling my own defence in court this time so I would not make the same mistakes as before. The only good thing that came out of it was that I was given bail again, albeit with conditions of residence and curfew and that I do not drive a car again whilst on bail. I should have been grateful really but I left the courthouse feeling slightly miffed at the judicial system and made my way down Corporation Street in town. All this was big news in the conspiracy, so obviously I had been 'broadcasting' all the time

THE DARK SIDE OF THE BBC. A DISTURBING TRUE STORY.

from my place in the dock and all of town were ready for me. When I thought again about dispensing of the services of Mr. Pugh and thus redeeming the validity of the trial by conducting my own defence, a comment was quickly made in support of this by a BBC girl as she passed me in the street. She grinned broadly and said "yes, I like that". Her comment connected to me just as I passed her and it made me smile to think that with all the brainwashing going on of the general public, people would ultimately still be people. An encouraging thought.

It was about 5pm when I rounded the corner near Hill Street to get to where I had left the car the night before. Who should I see strolling towards me on his way into town? I saw Des Mulcahy who I guessed must be on his way up town for a few bevvies. I enquired what he was doing in town and he told me that he had had a couple of pints with one of the local lap-dancers that worked in the vicinity. Then he asked me to go for a drink with him.

Now when Des asks me to go for a drink with him, a drink usually means an all day session so I had to tell him that I had no money. He told me that he would buy my drink for me but soon enough, I eventually borrowed £20 off him as I usually prefer to pay my own way. We went to the 'Briar Rose' first of all and I could tell that Des was well under the influence and revelling in it though I was still stone cold sober and still broadcasting to all around me. It was when I heard some-one say "let's see who he kills first" that I started to get irate at the proceedings but this same some-one soon disappeared so I became more care-free as the beer flowed. As it was, neither of us cared a damn as we enjoyed our exploits later on in the 'Square Peg' on Corporation Street but the name BBC cropped up occasionally around me by the women we chatted up, just to

THE DARK SIDE OF THE BBC. A DISTURBING TRUE STORY.

keep me in the picture I suppose. Both of us were capable fighters in our own right but together we didn't feel bothered by anyone enough to get too serious so when we were on form, we usually cleaned up in the romance stakes. The telepathic broadcasters didn't like this at all and tried to discourage any romance for me but I was too famous by now to be ignored so I still dated occasionally, though I was really having to save myself until after the expose. They were trying to starve me of all love and affection, hoping to hurt me that way. Des and I left the Square Peg well before closing time and made our way down Corporation Street in order to go home and I stopped off for a call of nature on the way. When I returned to where I had left Des, he was involved in a confrontation with some young Asian guy and it was about to get nasty. Nasty for the Asian guy that is. He was obviously out to cause us some harassment but when I envisaged the sort of damage we could do to him had we have wanted to, I just laughed at the fool and shooed him away with a brief description of what Des could do to him on his own. It was enough to convince him and he quickly fled the scene with his tail between his legs, which fortunately for him were both still intact. We just laughed as we made our way home to Small Heath.

    Not long after that, Des and I went for a night out up town again on Broad street. We ended up in our favourite pub/club for what would be our last occasion for some time, the Australian Bar or 'Walkabout'. We were taking the night nice and easy without any serious revelling or dancing and at one time we both sat at a table together just talking about things in general when I decided to go to the bar to get the drinks in as it was my round. As I waited at the bar to be served, I espied a pretty damsel waiting at the bar next to me. She was a

THE DARK SIDE OF THE BBC. A DISTURBING TRUE STORY.

blonde haired woman about the same age as me, maybe younger, and eventually the inevitable happened. Our eyes met, we clicked straight away and we started talking. It was so intimate that I began to feel a stirring in the old lower regions for a time so I decided I would have to take this further. We flirted for a few minutes before she returned to her party of girls who were based somewhere behind us and I returned to my table with Des. I told him about our meeting at the bar and of how she had caused a stirring within me and we both laughed about it for a time as not many girls have done that to me before. I told Des I would have to go and bump into her again, but within a few moments of saying this, she appeared out of the crowd and stood smiling before me. I called her over to me and we talked again for a short time before her friend came over and was not too happy about our liaison. I told the other girl to go away and leave her friend to talk with me alone as she was old enough to look after herself, which she did without too much reserve. I found out my little blonde friend's name was Sue and she lived close to me in Yardley. That was good enough for me to want to see her again so I asked her for her phone number which she gave me before she and her party left the club soon after that. Apparently, one of her pals was a bit drunk so they all left together to share a taxi home. I phoned her a couple of days later from my flat and we must have talked for over an hour before we arranged to meet again. The chemistry was rising. I arranged to meet her on the following Tuesday as it was payday for me so she gave me her address and I picked her up from her house at 8 o'clock that night in the car. We talked and laughed all the time before going for a McDonalds drive-through first of all whilst deciding where to go for the rest of the evening. I knew she was aware of who I was and what was

THE DARK SIDE OF THE BBC. A DISTURBING TRUE STORY.

going on in Birmingham around me so I thought I would just cut all the shit and talk openly about it. If she didn't like me then she wouldn't be here with me now would she? I even played her my tape in the car of when I was mentioned by 'Tushar', one of the dj's on Heart fm radio, who plugged my book for me. I knew she was impressed so I thought I would bring her home to my flat (the executive suite) and impress her more. We bought some drinks on the way from Ahmed's on Hobmoor Road and settled down in my flat around 9 o'clock. It was the night that Brave heart was on the TV so we watched that as well as listened to music. I showed her my membership of Amnesty International, I showed her clips from my book and other things relating to the battle against telepathic tyranny and how it would all soon end in court. She was well impressed with how I was committed to ending the situation here in Birmingham but she remained cool enough about it to maintain a subtlety about her. The chemistry remained between us throughout the evening until eventually I swept her off her feet and into my bed. We made love about four times that night although the drink did cause me to, shall we say 'lose concentration' a couple of times, which made us both giggle about it at the time until I was ready for the next session. We both enjoyed the evening together that night but it turned out to be our first and last night together in the conspiracy of things as events about to unfold would prevent any future for us, especially with the Evening Mail involved. Still, I enjoyed it.

THE DARK SIDE OF THE BBC. A DISTURBING TRUE STORY.

DRINK DRIVE NO. 4

I got up early as usual to go to court again at Birmingham Magistrates Court for three of the outstanding drink driving offences. I drove my car into town and parked it in the Evening Mail car park next to my old solicitor's offices in printing house street of all places. Piss take or what? I then proceeded to make my way

THE DARK SIDE OF THE BBC. A DISTURBING TRUE STORY.

to the courts. When I arrived there I met up with an old school buddy called Terry Adams. He was going as a witness at a trial against one of his neighbors who had seriously assaulted him one evening when he had been too drunk to defend himself. I, myself was in court 9 and Terry was in one of the courts upstairs so we went our separate ways and agreed to meet up later in which ever court we were in ended last. It turned out to be my court as Terry's trial ended first and he made his way down to my court. We waited together in the public seats for my solicitor to arrive from Carvers who were still officially representing me as I had not officially sacked them in order to conduct my own defense. A few solicitors from Carvers were there to represent me but I really only wanted Brian Pugh, as he was the more familiar with my defense procedure. He arrived eventually and although previously before, on another court hearing, I had given him four witness statements exposing all the telepathy going on in Birmingham by the BBC so far, Brian gave me the distinct impression that he was intending to have all my cases dropped in court without trial. I explained to him that that was not what I wanted at all but he exclaimed somewhat concerned "but it will help me!" I felt disheartened for both of us as I understood that he would feel the pressure of so big an expose. I finally decided then that I really would go ahead without legal representation this time and conduct my own defense again. I was more than capable anyway. We both took up our positions in the court-room and I stood before the two magistrates. I stood in the dock by myself for a couple of minutes sizing up the 1 male and 1 female magistrates when it slowly dawned on me who the female magistrate was. She was the same one presiding over my very first trial which was for shoplifting, when she had so ruthlessly instigated the cover-up by ordering the

THE DARK SIDE OF THE BBC. A DISTURBING TRUE STORY.

psychiatric report made on me. As this was a major event she had instigated in the conspiracy to cover up and because I was still thought and sight broadcasting, she became very nervous when it became obvious to me in my mind who she actually was. She herself could not stand the pressure of me knowing who she actually was and so when she realized that she was recognized, she got up from her seat of power after explaining to the clerk to the court that she had previously adjudicated at another court case of mine and so could not sit in on this one. Had I not recognized her, she would have stayed in her seat and took part in another cover up in this case also. As it happened, the male magistrate was left to adjudicate by himself, so my brief addressed him with the basis of the course of action he would be taking concerning my three impending trials. I understood, as anticipated, that the psychiatric report ordered at my first trial would now be linked to all other impending trials so that the said report could then infect them all together. When he had finished his patter, the magistrate himself then explained to me what was happening as if he was talking to an idiot. I simply showed no emotion or reaction to what had unfolded in front of me as I left the dock by myself and then the courtroom. As Terry walked ahead of me, I stopped for a few seconds to talk to my solicitor. As I already knew that I would not be using him anymore after all, I simply managed to smile at him when he explained to me that he would be getting all the charges dropped against me. Under normal circumstances, this would be seen as a good result but in my circumstances, it meant that all the previous work I had done and been through would have been wasted and the chance of a normal future still not yet a reality. We shook hands and I left Brian with a smile of resignation on his face as he contemplated what he had to do and I

THE DARK SIDE OF THE BBC. A DISTURBING TRUE STORY.

went off to catch up with Terry. I even explained to Terry that I would be conducting my own defense from now on as we walked back to the car, which mildly amused him as well. When we arrived at the car park, it turned out that my car had been clamped by some clod from the Evening Mail and I had no tools on me to force it off. A little dismayed and a little amused, we went in to the Evening Mail building to sort it out. I was directed to the security offices on ground level inside the building where I was met by a security guard on duty at the time. I explained to him what had happened to my car and he too seemed a little nervous as well as he explained to me that he would have to make a phone call in order to sort it out as there was a fixed penalty charge of £30 to all un authorized cars on the car park and I had no money on me at the time. I knew the Mail were involved with what was going on around me so it amused me to see all the confusion I had created just by entering the building to such an extent that some of the female employees from the offices upstairs who had come in behind me, now walked passed me with bright shiny eyes on their faces and the usual smile that went with it, just like the BBC. I waited for the security guard to come back for a few minutes as even Terry was now smiling as he watched what went on, until soon enough the guard returned to me with some bad news. Apparently, I was told that not due to himself but due to the big boys upstairs, the charge could not be lifted and I would have to pay the £30. I told him I would come back tomorrow with the money. I left Terry at the bus stop and went home by myself. This was on a Thursday and I realized that I wasn't getting paid until Tuesday the following week. I had to return to the Evening Mail building the next day in order to explain this to them as I didn't know if the charge would be increased if the vehicle was left there for any

THE DARK SIDE OF THE BBC. A DISTURBING TRUE STORY.

length of time. The security guard on duty on this occasion relayed my situation to the big boys upstairs and within a few minutes the head of all security came down to deal with it. We stood there together face to face looking directly at each other and this man came knowing he was going to be looking directly into a camera, showing himself telepathically to all the millions of people around Birmingham at least, so we both had to be strong. I explained the situation to him myself and I even blagged him a bit by saying that I worked for the Evening Mail, "well, sort of". I detected that he was amused at this so it made me smile myself. As soon as this happened he turned and said the word "there" to the guard, indicating to the broadcasters in the studios that he wanted the broadcasting to be cut there with me with a smile on my face. Then he told the guard to sort it out so that the clamps were removed from my car with no charge to myself. He let me off without paying.

The guard then followed his instructions within a few minutes and my car was released from its shackle. As I had cycled into town on my bike, I hadn't brought the keys to the car with me so I had to cycle back out of town again in order to retrieve them. I steamed out of town at full speed on my racing bike and as I passed a pedestrian on the street en route, he seemed to lean into my peripheral vision and exclaim aloud the word "mad". He did this to indicate to me and everyone else that I was mad to smile like that towards the Evening Mail, but it didn't bother me at all as I had saved thirty quid out of it for myself. I cycled at top speed all the way to Small Heath until I reached Monica road, then I got off my bike and took a walk when I neared the steep hill at the bottom of the road. After a short walk, I bumped into a mate of mine from the area also called Tony, a black guy with dreadlocks, except that today I saw him

THE DARK SIDE OF THE BBC. A DISTURBING TRUE STORY.

for the first time without his locks. He had shaved them off. We laughed about it for a while before we both decided to go up the town for a few pints while I collected my car from the Evening Mail car park. I dropped my bike off at home and retrieved the car keys at the same time. We jumped a bus into town and ended up in the square peg on Corporation street where we seemed to slowly become the centre of attention. I was used to it of course by now so remained cautious and shy until there must have been about six guys around us at our table who were genuine enough to warrant a good time to be had. One of the guys betted me that one of the two girls sitting at the table next to us was not wearing any knickers so without further ado, I put the record straight by marching over to the table where they sat and asked her point blank if she was wearing any knickers. Of course I had my cheeky grin on my face as I asked her so she smiled back and told me that she did. We laughed about it as I had won the bet and the banter carried on now with all of us involved when I returned to my table. We were all having such a good time until someone appeared and quite literally tried to throw a spanner in the works. I noticed him standing beside me looking rather menacingly which made me react to him in the same way. I felt I knew he was trying to intimidate me so I leaned in his direction and asked him openly what the matter was. He had made himself stand out like a sore thumb from the rest of the others but when I asked him what the matter was he seemed to relax his menace and began to tell me how he had been stood up by his mates and was all on his own. I didn't really believe him but invited him to join our table anyway and enjoy himself. He didn't sit with us lads but opted to sit more with the two girls that were with us now which some of us thought was a bit ignorant but if he wanted to steam

THE DARK SIDE OF THE BBC. A DISTURBING TRUE STORY.

into the chics then that was up to him. It didn't bother me. It was when he introduced himself as 'Spanner' that we began to see the connection. He didn't laugh much either so when other Tony took the piss out of him by saying he looked more like a screwdriver, it sort of clicked then. Mr. Spanner soon enough got up and shot off to the toilets which the two girls took as an opportunity to leave him so when he returned to the table, they were gone. He enquired as to their whereabouts and I replied that they had escaped from him on purpose. I invited him to sit with us lads but I think he couldn't handle the banter for some reason so he seemed to quickly neck his pint and shot off. I shook his hand before he left just to let him know we enjoyed the craic and then he was gone. Soon after that the rest of the lads with us seemed to talk with admiration about me so being as I had still yet to drive the car home from town, I decided to leave the company along with other Tony on a good note and head off home. We shook hands as we left and walked off towards the doors in order to leave. That was the last thing I did before once again, my mind was blacked out by the studio operators of the dark side of television. A truly evil and dangerous thing to do to someone about to drive a motor vehicle, especially when he is carrying a passenger. Anyway, I was kept in a state of pitch darkness all night as I continued to drink alcohol without even knowing about it, while the studio operators had some fun at my expense. I was left in total darkness long enough to get drunk and crazy enough to warrant the game "who does Tony want to fight?" to begin. I was being used like a toy again as the first face to come to light was suddenly flashed in front of me to see how I would react. The first face I saw was that of a black guy's who I worked for called Ken. He owed me and Des some money in lieu of work we had done and

THE DARK SIDE OF THE BBC. A DISTURBING TRUE STORY.

he hadn't paid us so we had stopped working in protest until we were paid. Fortunately I was still quite sober at this stage of my 'awakenings' so I handled the situation quite well. So well in fact that I managed to get the £100 we were owed after a little conversation was held outside. I must have impressed Ken enough for him to have paid me after he had waited before me in anticipation of my 'awakening'. Then I was completely blacked out again. Some time must have elapsed and more alcohol drunk before the next intended scenario took place. The next face to come to light was that of Des. He didn't look too happy at the time of my 'awakening' so I guess that he was being used against me at this point as well, though I know Des is normally ok with me. He suddenly appeared in the light and it was only a matter of seconds really before I was blacked out again but I managed to tell him the score of how I had retrieved the money that was owed us and managed to give him £50 that was his half. I was left to more darkness again then until what must have been towards the end of the night when I was brought back into consciousness again for what was to be the very last time. I must have been quite drunk at this stage because when I was brought to my senses, I was standing motionless in the middle of the pub, which was the Samson and Lion. I saw Tony's face, the same one I had been with all day and he sat quietly in his chair looking towards me looking very, very sad. This was the first time I saw him since we were leaving the Square Peg and I don't think he was too happy about the day's events. I was blacked out again into unconsciousness and the next time I came to, I was waking up in Stechford Police Station after the Police had landed on my case yet again to round off the days events. I was not to be given bail at all this time but to face the prospect of a long time behind bars where my strength

THE DARK SIDE OF THE BBC. A DISTURBING TRUE STORY.

and resolve to fight against the misuse of telepathy enforced upon me could be put under extreme pressure, until I was broken.

It was Friday evening 8th March when I was arrested and I spent the whole weekend in Police cells waiting for court on Monday 11th. I was left thought broadcasting continuously for the duration of my time in custody but it was the beginning of my time spent incarcerated that was the most revealing. I was at the mercy of the powers that be so there was no escaping whatever lay ahead of me, I would just have to deal with it. The Police were on my case first obviously and they worked around the clock in shifts at their leisure to harass my every thought and desire to sleep. I tried not to let it bother me too much as it was nothing new to me but it continued throughout the weekend until the Police, realizing they were ineffective against me, began to come clean and eventually told me who was responsible for all the telepathy in Birmingham at present. It was not just the BBC as I had always believed but they were now telling me that it was the Evening Mail Newspaper that was connected to television and telepathy as well. I had always known the Evening Mail was involved but now it began to connect that they were indeed telepathic broadcasters with BBC equipment as well. The whole Birmingham versus London thing began to fall into context. But I still didn't believe too strongly as I knew it was all about Television, as in the BBC. I would learn more from people as time went by that would show that the Mail really was involved in some way and my suspicions of years since were well founded.

ONE FLEW OVER THE COOKOO'S NEST.

THE DARK SIDE OF THE BBC. A DISTURBING TRUE STORY.

On Monday 11th March 2002 I was transferred from Stechford Police station to the lock-ups underneath the Magistrates Court in central Birmingham. By now I knew I was thought broadcasting all the time so I became cool and defensive in my manner. It seemed pointless to use the services of Carvers Solicitors again as it was clear that they would not be helping me in my cause so I decided to represent myself in court again. I had not intended to expose the existence and misuse of telepathy in court with silly charges of excess alcohol but now I had no option but to tell the truth anyway and let the justice system take its course as one would expect. The security services that operated for the efficient running of the lock-up's were friendly towards me to begin with and made it obvious to me that I was somehow famous to such an extent that they advised me that it would be in my best interests to hire the services of a solicitor, especially in view of the fact that I would be making a bail application upon my appearance in court. This seemed like good sense to me at the time so I decided to use the services of the duty solicitor that day knowing that I could always sack him after I got bail, if and when he betrayed me. The duty solicitor that day was a Mr. Reece but I don't know what firm he was from. I was taken to see him after a couple of hours waiting in the cells and we introduced ourselves in the interview room under the courts when we were left alone together. I had no intention of avoiding the issue so I came straight to the point and told him of my intention of exposing the root of all my problems, the existence and misuse of telepathy being used as a weapon against me. He listened without interrupting as I spoke and when I finished, I paused to await his reply. The first thing he asked was to enquire as to whether I had ever been

THE DARK SIDE OF THE BBC. A DISTURBING TRUE STORY.

treated in a psychiatric hospital. His intentions then became obvious to me and it caused me to lose my temper at his insinuations that would seemingly lead to another cover-up. I snapped back at him that he should not get involved with all that bull-shit and that he should at least now go into court and represent me with my application for bail. He was taken aback a bit by the seriousness of my little outburst and realizing that I was not going to be fobbed off with another psychiatric cover-up, he had a change of heart. I could tell from his body language that he was now intent on helping me in my plight. He left to take his place in court and I was returned to my cell. I didn't have to wait too long before I was called to take my turn in court and was soon escorted to my place in the dock. The court had been in session all morning near enough before my case was called but for some reason, when I was shown into the dock, the presiding magistrate was not in attendance in court. He was in his chambers awaiting my arrival I suspect, probably so that he could make his own dramatic entrance into court like some kind of movie star. Well let's face it, everyone knew I was a living TV camera and so acted accordingly, especially when they knew I was 'filming' an important scene. He made his way to his seat at the bench by himself but my attention was drawn to the female prosecutor who stood next to my solicitor a few yards away. I saw her lean intimidatingly towards my solicitor and thrust some paperwork before him, using her index finger to point to something written in the document. I could plainly see that it was my previous convictions she was drawing his attention to and her finger was pointing out my last conviction some ten years ago when I was sentenced to a hospital order. She was in fact trying to impress upon my solicitor that there would be no expose of the BBC and their counterparts if she could help it

THE DARK SIDE OF THE BBC. A DISTURBING TRUE STORY.

and that it was all heading towards another psychiatric cover-up. I felt my jaws tighten. My solicitor seemed to deflate a little at the sight of a hospital order on my record and the scandalous implications it evoked. He resigned himself to failure there and then and I was left at the mercy of the presiding magistrate. Although I was not confident myself of getting bail as this was my fourth offence of a similar nature within as many months, I did believe that I would possibly get bail on the grounds of the human rights issues that were so obviously involved. Imagine going to prison with your innermost private thoughts and emotions being broadcasted to everyone around you and with everyone invited to intimidate and harass you because of those private thoughts and emotions. Not something to look forward to obviously but within a matter of minutes within this court-room, this was to be my immediate fate. I was refused bail and returned to the cells underneath in order to await transfer to a secure prison for one week until my next court appearance, at which time I would be allowed to make one last and final bail application. They thought they were sending me to hell. It all started as soon as I was returned to my cell. The security guards seemed to be the instigators of the harassment that ensued as they took it in turns to stand just out of sight outside the cell door and bombard me with derogative and insulting comments in accordance with whatever thoughts were being transmitted from my mind. They also brought other prisoners to my door and allowed them to continue the harassment outside it as I awaited transfer to prison. Then there was the voices that were broadcast from the studio that I was led to believe were established within the offices of the Birmingham Evening Mail. They were transmitted into the ventilation system that ran above each cell and were concentrated around the

THE DARK SIDE OF THE BBC. A DISTURBING TRUE STORY.

vent opening in my cell that I shared with three other detainees. I was unfazed by all this that was going on and even laughed to one of my cell-mates as I asked him if he could hear the voices in the vent system that were the Evening Mail. He smiled dryly and replied "I can hear something". My lack of concern had a positive effect on one of the studio broadcasters who then transmitted the words "And he comes up". He was saying to myself and the audience that I would eventually come up to the offices of the Evening Mail and get things sorted. These words seemed to have a discouraging effect on the others as the 'voices' began to dissipate into quiet after that but the involvement of a local newspaper embroiled in criminal activity which involved the misuse of telepathy was now becoming more apparent.

It was very late in the afternoon when the guards finally decided that they had had enough fun that day at my expense and that I wasn't going to shed any tears about it so they began loading up the van with prisoners bound for Winson Green Prison, Birmingham's local Prison. There was not enough room in 'the Green' to cater for all the prisoners detained that day so the guards thought they would have one last piss-take before I was gone. They took me out of my cell and escorted me in handcuffs onto the Prison van, then just before I was about to sit down in the enclosed cell in the van, which was actually smaller than a single wardrobe, they told me in front of all the other prisoners that I had to get off the van now. When I enquired as to why I had to get off the van now as I could tell something fishy was going on, I was told by one of the women guards "because Birmingham doesn't want you".
I felt really, really angry at that sort of snide comment coming from someone supposed to be in authority but my cuffs prevented me from dishing out some

THE DARK SIDE OF THE BBC. A DISTURBING TRUE STORY.

justifiable rhetoric onto the end of her nose so I just screamed back at her and the other prisoners in case they were involved "you mean I don't want fucking Birmingham". I had to let them know that I wasn't going to be creeping to Birmingham in any case even though I was a Brummie myself and that if there was still to be any Birmingham people involved against me in my battle against the evil media giants, then they would fall too. I was in Prison surrounded by criminals for God sake. No place for Mr. Nice Guy. I was taken back to the cells under the Court-house and the van departed for Winson Green Prison without me. I didn't get sent off to any other prison that day as I expected I would, instead I and a few others who remained were kept overnight in Steelhouse lane Police Station where I was allowed to make one phone call. I phoned my Mom to inform her that I wouldn't be coming home for a while but I got the impression that she already knew of my situation so I realized that the news was already broadcasted telepathically to and around Birmingham. One of the Policemen acting on duty that evening asked me if I had enough tobacco to last me until the morning and when I told him that I didn't smoke, he was so surprised and taken aback at this that he said " you're about the only one that doesn't!" He seemed to come to realize that I would probably come through my period of incarceration as I was seemingly more healthier than most of the other prisoners, a fact that would serve me well in the presence of the 'grapevine' where word passed from mouth to mouth. I eventually went to sleep on a hard bench that night amid Police 'voices' and awoke for breakfast the next morning with an even stiffer back. We were kept hanging around all day before news was given to us that we were not going to Winson Green Prison as anticipated but that we were going off further up north to Leicester

THE DARK SIDE OF THE BBC. A DISTURBING TRUE STORY.

Prison some 30/40 miles away until my next appearance in Court on Monday 18th March '02, which was now six days away. We were termed as 'lock-outs' but I guessed I was being set-up to see how I would get by in a Leicester Prison with what they would be told was with the whole of Birmingham and his uncle on my case. I wasn't really concerned in the least because I knew I always had a fighters chance of getting by whether it was a Brummie boy or a Leicester lad I came up against. And so, undaunted by my predicament, we set off for Leicester Prison late on in the afternoon that day with the BBC's radio 1 playing loudly in the background. These DJ's seemed smug and patronizing of me now in their manner as we travelled on up North with my thoughts and my emotional state being broadcast to all around me, probably because they were of the opinion that I was not going to mentally survive my period of incarceration. That made me smile. A girl on the radio argued in my defense and that made me smile too but obviously for a different reason. We arrived at Leicester Prison about an hour later and as the van drew into the Prison compound, I could see, as anticipated, that the whole complex was waiting for my impending arrival. I could see a scattering of Prison Guards around me looking nervously down at their shoes as my thought and vision was transferred into their minds and there they saw real-life images of themselves. This was probably the first time they had witnessed this experience of telepathy, even though it had been going on some miles away in Birmingham for over a decade. It amused me slightly to see their reaction to a telepathic presence in their Prison but I knew it could turn nasty at anytime so in order to let them all know that I was not in the least bit worried about my predicament, I transmitted a message to them all telling them in not so many words to lock up their wives

THE DARK SIDE OF THE BBC. A DISTURBING TRUE STORY.

and daughters as Tony the telepath had arrived in Leicester Prison, intent on rape and pillage. I was only joking of course but I got the message across, that because I was basically an unwilling extension of television, it did not make me weaker in any way and that I was quite capable of taking on the threat of any Prison regime and that I could stand my ground against anyone within that regime. It seemed to do the trick because when we alighted from the van and adorned the reception area with our presence, the attitude towards me from all the others was that of mild humor mixed in with the desire to stay clear of me if at all possible. That suited me fine. The scene was set.

We had to go through an induction period first of all before we could join the main prison population, as you do when first arriving at Prison. We were kept on a separate landing from the others as we had to see the doctor, the probation etc in order to acustomize ourselves to the rigors and rituals of Prison life. When I saw the Doctor upon my arrival at the prison, I found him to be sly and obnoxious to begin with as he seemed to think he could talk to me as though I was blissfully ignorant of what was going on around me and that I was, in his eyes, a lost cause. He was to me another typical example of the way some people would side with the big bad guy out of cowardice, rather than favor the underdog. He obviously didn't know that I had been kicking ass for years. Well, they wouldn't tell him that would they? They would never fool any more new recruits if they did!

I dispensed myself of his services and told him that if I didn't see him through the week, I would see him through the window. Next I had to see the probation service and found myself being interviewed by a girl called Emma. I really did

THE DARK SIDE OF THE BBC. A DISTURBING TRUE STORY.

need to get a probation report done for my next bail application at court so I asked Emma if she would prepare one for me in readiness of my next Court appearance. She told me she would have one ready at my next bail app' so I relaxed a little at the thought of maybe being granted bail next time. I detected that she was aware of telepathy because I engineered a few nice and sexy thoughts about her just to gauge her reaction and she reacted perfectly to my little test. She probed me for information as well so I just had to inform her that I was in the process of writing a book called 'the dark side of television'. I knew that news of this would soon spread and that people in authority intending to 'take up arms' against me would then have to seriously think twice about their actions. This was another weapon I had installed in my arsenal in preparation of things to come. The verbal existence of my book was not enough to sway Emma to my side on this occasion though it seemed as she didn't even produce the probation report for me in court as requested. I was obviously up against a ruthless and nasty piece of work as my opposition but if the BBC and their counterparts are a big, big tree then I am a small axe, ready to cut them down.

The whole prison was on my case it seemed from the off as I was the new fish in the market and telepathic with it to boot so I was targeted for harassment from the word go. I could hear some of the other prisoners in the main population leering and jeering at my private thoughts from the safety of their obscure positions in an attempt to frighten me into submission even before I had joined the main population. I couldn't wait to get over there now just to get in amongst the big brave troublemakers but after I was eventually sent over there into the lion's den, the lion's had turned into little pussycats. I

THE DARK SIDE OF THE BBC. A DISTURBING TRUE STORY.

spread paranoia about the main Prison as I walked the landings to and from meal times with an air of indifference and cool confidence about me which most people couldn't seem to come to terms with. They couldn't comprehend how I could be so cool and confident considering my predicament which unnerved them a lot really to such an extent that most of the prisoners began to take my side in the evil conspiracy of things that consumed us all. My tactics were to push things to the limit in order to see how much danger there was in any repercussions for my actions. I was genuinely unafraid of the consequences because the worst that could happen would be that I would get involved in a fight and knowing that that was probably my worse case scenario some months ago, I had invested in a built-in gym inside my flat for the long run. Win or lose, I was capable of fighting any-one. Well, I might be tempted to avoid the likes of Lennox Lewis and Mike Tyson but the likes of Danny De Vito would be a walk in the park for a man of my caliber. (Laughs).

I decided to force the hand of the enemy within and so one night after we were locked up for the evening, I began to continue writing 'the dark-side of television' in my cell with pen and paper. I knew that I would be broadcasting my actions to the entire prison population and that if there were to be any repercussions about it, they would take place the next morning en route to or from the servery at breakfast time. The fact that I was a bit of an adrenalin junkie that thrived on pressure and danger made things easier for me I suppose so I endeavored to continue to write my book in my cell that night even with the presence of the 'voices' in my head that tried to deflect me from writing anything too risqué in their opinion by trying to intimidate me when I did. Anyone who had to hide their voice in my head to intimidate me because they

THE DARK SIDE OF THE BBC. A DISTURBING TRUE STORY.

couldn't do it face to face was in no way going to intimidate me at all. I just ignored them usually though I did analyze them occasionally just to keep abreast of what they were up to even though I didn't really give a damn. I went to breakfast the next morning with an air of expectancy about me though I suppose I did emit a cool, menacing aura as well. As you do. I collected a tray from the servery and joined the queue to collect my breakfast knowing full well that they were all tuned in to my emotional state and were waiting like a pack of hyenas for any sign of a weakness. I was buzzing off the intensity of it all and was ready to explode onto any form of threat I might experience. Confidence kicked in again as it always does in do or die situations and I began to enjoy the thrill I got from ruling them all in the face of adversity. It was probably comparable to a sky-dive from 10,000ft I suppose though what the fuck a sky-dive from 10,000ft feels like is beyond me at the moment. I reached the far end of the servery in the slow moving queue without a word being spoken so I knew that one of the last two servers on the other side of the counter would be the one to make some kind of comment that would make or break the situation. I kept my cool composure as I approached them although my jaws did seem to feel as though they were locked together in anticipation of what one of these lads would be feeling pressured into saying now. The guy who was second from the end served me first obviously and as he did so he made no bones about what he wanted to say. He called out the words "back-up", which implied that the Leicester firm would back me up against those nasty little posers from Birmingham that were seemingly on my case, spreading poisons in my wake as I advanced. These people were only capable of back-stabbing so they were not taken very seriously by many especially when it was

THE DARK SIDE OF THE BBC. A DISTURBING TRUE STORY.

obvious to see that I certainly held no fear of them. After that little episode I really wasn't too bothered about what happened from then on because I knew that in the event of a serious situation, people who were not controlled by 'voices' were more likely to show support for me than to show adversity. They probably clicked with my cool sense of confidence under pressure which was an indication of my ability to deal with anything that could be thrown at me. I would soon have to show them all that this confidence was not born out of a falsity as the pressure was soon to be intensified against me. I was padded up with an Asian lad from Birmingham would you believe who had been in the prison for some time who was either actually doing or looking at a long sentence if convicted. His name was Safrash and he was on the ball straight away as soon as I walked into his cell that first time. He had the radio on and as I walked in to the sounds of some local Asian station, I heard someone on the radio say " let's just say it's those media people". This made us both laugh some as we both occasionally exclaimed aloud to each other "those media mutherfuckers", "those media motherfuckers". We both knew it was the Evening Mail and the BBC that were still involved but I was never in the mood for discussing it with anyone that I couldn't benefit from so we left it at that initially. We had some mad discussions on religion and the meaning of life and so on because this guy could talk some especially now that he was famous in the prison with Tony as his cell mate. He was ok really though considering he was a Brummie and so liable to be lured into operating against me but he seemed capable enough of acting on his own initiative and so we managed to get on. The local radio station that was almost constantly on seemed to always make positive comments as well so that was another feel-good factor. I had to

THE DARK SIDE OF THE BBC. A DISTURBING TRUE STORY.

see the Doctor shortly after my arrival in order to get some cream for a little rash I had on my finger and when I turned up at the appointed time to see him, it turned out to be that same Doctor I had seen on my induction period a few days ago. This time however, he was not alone. He was accompanied in a professional capacity by a very pretty little Chinese nurse who I soon discovered was called Paula. I sat down to talk to the two of them but when she was introduced as a psychiatric nurse I was a little suspicious to say the least and knew that I would have to be careful of what I said to her at that meeting. I still had four trials pending in the magistrate court that could expose the existence of telepathic rule that still operated illegally in Birmingham and the control it had over my life in general. I didn't want Paula to get involved in another cover-up so I remained non committed to some of the questions that were asked of me even though I got the distinct impression that she was not there to get me nutted off to the hospital wing but was just curiously interested in how I operated as a person at the centre of such seriously heavy shit. When I told her that I would be going back to court again soon this coming Monday, she replied "yes you will have to go back and face Birmingham". She was laughing as she said this to me so just to piss on a little fire, I said to her "you mean Birmingham will have to face me!" I was laughing as well when I said that which just went to show how bothered I was about what the future might hold for me but it obviously unnerved these two bright little sparks who changed their attitude from patronizing of me to that of only praise and encouragement for me. Paula even commented on how I was a big lad and capable of taking care of myself in any event. The Doctor seemed more anxious at the fact that he had believed me to be a soft touch destined for the long

THE DARK SIDE OF THE BBC. A DISTURBING TRUE STORY.

term confines of some seedy psychiatric institution. I also let them know that I was writing a book about the dark side of television as I supposed that they knew by now anyway but I didn't inform them as to whether it was a true story or just fiction. I enjoyed playing cat and mouse with them simply because Paula was so pretty for a Chinese girl but I began to have serious reservations when she asked to see me again before I left for Court on Monday. I told her I would think about it but she basically tried to force a Sunday meeting in my cell on me as I was leaving which again I told her I would think about but the last thing I heard her say as I disappeared out of the surgery was "see you Sunday then." This forthcoming event played on my mind for the duration of the rest of the week as I was unsure if I was allowing the beginnings of a cover-up to take place for the sake of seeing a pretty face or if I was allowing her to express some genuine interest in my predicament to me for the sake of seeing a pretty face. I decided not to see her in the end and would deny her access to my cell if she did turn up on Sunday. As it happened I relented to myself by the time Sunday came around and I wasn't really bothered about seeing her or not. With my inner thoughts always subjected to constant broadcasting, this was picked up on and a situation engineered whereby Safrash would be allowed to leave his cell by himself for a while leaving me alone by myself and then hey presto! Paula would just happen to be passing by my cell. This made me chuckle to myself when I saw what took place so I allowed her to come in just for some much needed entertainment if anything. I learned from miss pretty that she was actually involved with me because she had been written to by someone from my old friend the Medical Profession outside. I think she said it was Mick O'Hanlon that had written to her enquiring as to my situation. Mick was my

THE DARK SIDE OF THE BBC. A DISTURBING TRUE STORY.

social worker assigned to me after the charade had begun that was the medical cover-up but he was not directly involved in any of my periods of incarceration himself so I allowed him to continue to function for the present without any ill-feeling as he was just doing the job assigned to him I suspect. He obviously knew what was going on in my life but there wasn't much he could do to intervene. The task of officially exposing the existence and misuse of telepathy in Birmingham today was laid firmly at the feet of the only person who cared enough to do it. The victim. Me.

Paula and I talked openly for some time about the BBC and so on although I was prone to answer "no comment" to some of her questions in order to remain none committed which was amusing at times. When she was ready to leave she impressed upon me and whoever was listening in that all I wanted to do really was to just get my life back together which was easy for her to say but I knew I was in for the long haul basically. I knew I had to free myself from these electrical 'chains' that were an abomination of the dark side of television to which I was connected to but I also knew that that wasn't going to happen until after I had sued the likes of the BBC in the Civil Courts as there was no way they were going to leave me alone until then when it would be game over. As someone on the TV from the BBC told me ...... It aint over 'till it's over.

The involvement of the Evening Mail Newspaper had no effect on the task in hand. I often wondered now whether my oldest rival Robert Maxwell was really with Central TV all along or if he was really involved with this kind of illegal broadcasting with the likes of the Evening Mail. I had been told on numerous occasions that an official investigation would have to be called for eventually to determine this for legal reasons but their involvement was not so important to

THE DARK SIDE OF THE BBC. A DISTURBING TRUE STORY.

me as all would fall into place once I had exposed the real instigators of all the telepathy in Birmingham, those who still caused it to continue to this day and those who would soon be held accountable to bring about the end of it. The BBC.

It was Monday 18th March 2002 when I awoke the next day, the day of my second Court appearance whilst in legal custody. I was due to make my second and final bail application that day and it occurred to me that the powers that be may have only intended for me to be locked up for one week in prison just to teach me a lesson or give me a scare even, in which case I could be going home within a matter of hours. It would have been nice I guess but I was not banking on it because I knew that I had four trials pending which could kick-start the legal process that would bring about the end of telepathy used as a weapon against ordinary people like myself and so I was half expecting a case-load of shit to be thrown at me in the run up to those trials. Whatever the case, I had to go through the motions anyway because of the legal process, so I would soon see it all for myself in due course. The van that ferried us arrived at the Courts on time near enough and I was soon taken inside in hand-cuffs to my allocated cell. As I was standing outside my cell ready to be searched after the cuffs were removed, I looked upon the figure of a man approaching me from the distance who was clad in a well fitted suit. As he drew nearer to me I vaguely recognized his face from before somewhere and as he realized that he was being scrutinized he said "alright" to me as he passed. I knew that I knew him from somewhere but it wasn't until I was

THE DARK SIDE OF THE BBC. A DISTURBING TRUE STORY.

locked in my cell that it dawned on me who he was. It was my new solicitor that represented me in Court the previous week, Mr. Reece.

It confused me a little at first as to why he had not stopped to talk with me when he saw me outside, after all I was telepathic and so he obviously knew that I was coming. Then when you consider that he was also supposed to be representing me in court that morning, I sort of got the impression that all was not right. The hours passed by instead of minutes and he still hadn't even come to see me as all solicitors normally do when their client is in custody. I am not a fool, as some people may have been led to believe, so I soon began to get impatient at all the waiting I was doing for some-one who I now began to suspect was not even going to show. I asked the guards to make enquiries for me as to the whereabouts of my brief but they just kept saying that he would be here soon. It was ridiculous really, I knew in my heart that this was a piss-take and that my solicitor was not coming which made me more and more irate but when I reflected on the fact that it was every British citizen's fundamental right to legal representation in court, I had difficulty believing that this was really happening. Then when he still hadn't shown up by mid afternoon, I became more or less convinced that this was indeed a reality. My mind began to search and analyze everything that took place the previous week concerning my new solicitor. And then it hit me. Before a solicitor will represent his client in court, the client will normally have to sign the 'green form' first, which is a sort of contract, otherwise the solicitor will not get paid for his work by the legal aid system. You've guessed it! I had not signed a 'green form' with Mr. Reece and so he was not legally bound to represent me in court as there was no proof of a contract. The fact that he had represented me in court the previous week was

THE DARK SIDE OF THE BBC. A DISTURBING TRUE STORY.

of some merit to him I suppose as he would not be getting paid for it. My immediate concern was the fact that I was due up in court without any representation at any time now and what was I to do about it. After a few loud screams, a few curses and a few boots to the door that held me captive, I quickly decided that I would conduct my own defense in court and would also have to make my own bail application. My cell-mates agreed with me, probably just to keep me calm and so, soon after I informed the guards of my intentions, I found myself alone in the dock but ready for action. I was not at all confident of getting bail but I had to go through the motions anyway and also, I just wanted to show the bastards now that I was not afraid to do it. I was in court no. 1 that day which is one of the two biggest courts in the whole building, probably selected at the time to make me feel intimidated in my quest for justice. My surroundings meant little to me, though as I looked around this vast, empty arena, I did somehow feel a little small and insignificant for a second before I saw a familiar face looking back at me from the public seats. It was the only face in the public gallery. It was Des Mulcahy.

It made me smile to see that he had come to see how I would get on in court this day, so I felt a bit happier from then on even though I suspected that I would be remaining in prison. I also looked around for a representative from the probation service who should have been there to produce the probation report on me as done by Emma from Leicester Prison but there was no-one to be found. The words 'stitched up like a kipper' sprung to mind. I wasn't bothered. I knew I was going to win in the end and that was more important. The magistrates sized me up as I stood alone in the dock before them and the male magistrate enquired of me as to if I was being represented that morning.

THE DARK SIDE OF THE BBC. A DISTURBING TRUE STORY.

The question was so incredulous that I had to smile back at his dead-pan delivery before answering him. I looked around me and replied that I was supposed to be but it certainly didn't look like it now so I would have to represent myself. Des looked on from behind me with interest at the proceedings but I think he knew that it wasn't just the BBC I was up against now but the whole establishment. I was at a huge disadvantage as well because I was the accused at this point in the eyes of the law and I was also currently detained in custody because of it so it was me that was deemed as the bad guy now. I had to admit that I was 10 per cent responsible myself for allowing myself to be ensnared the way I was due to my social drinking habits. It seemed I was to be frowned upon by certain studio based cowards when I appeared to be enjoying myself, especially when alcohol was involved. The fact that most people enjoyed my exploits when drink was involved made no difference whatsoever. My life was not supposed to be a happy one apparently, according to the dirty DJ's. These dick-heads were more amusing than me though in reality as far as I was concerned and they weren't trying to be.

My theory that I might just have to spend a week in custody in order to give me a little scare was soon blown to pieces when the Magistrate refused to grant me bail again and I was cuffed once more and ushered back to the cells underneath. I bid Des a farewell on my way back, doing well to contain my disappointment and anger at the proceedings as I did so and told him that I would write to him soon to let him know what was happening. I returned to my cell some moments later knowing that I was in for the duration now. I would be staying locked up until the day of my trial when I hoped all would then begin to resolve itself. This was my guiding light at the end of the tunnel. I was packed

THE DARK SIDE OF THE BBC. A DISTURBING TRUE STORY.

off to Winson Green Prison in Birmingham this time so now it was time to sort out the credibility of the wannabe gangsters I might encounter that may oppose me. There was no way I was going to get beaten into submission so I was quite prepared to stand up for my rights and fight my way out of this shit-hole for the duration, no matter how long it took. Win or lose, I didn't care.

It wasn't too bad in Winson Green Prison this second time around really, considering. The fact that I was on remand and not convicted was a bonus in itself because it meant that I didn't have to work and so I could spend most of the time in my cell like the others. When I was on the induction wing to begin with, C1, it was a bit annoying because most of the other con's were usually hanging around outside my door making comments on my thoughts and actions and I was a bit narked about that especially when I began to realize that most of them were cleaners from other landings who had gone out of their way to descend on the induction landing in the basement floor of C-Wing, just to be near me. They seemed to be really enjoying themselves mouthing off to someone behind a closed door, I could hear them laughing and giggling like school children through-out the first day, but when I was moved upstairs on the mainstream, things were a little different. The mouthing off stopped completely and things were a lot quieter from then on for a while, especially after I had faced my first cell-mate. I was located upstairs on the third landing, C3, and as I entered the cell he occupied, he was lying down on his bed with his arms behind his head. He stared at me for a second when I entered and I returned the look myself in order to check out the intentions of my new cell-mate. He was a big lad (wouldn't you know), so I had to let him know that he was not about to be two'd up with some muppet that he could take advantage of. A lot

THE DARK SIDE OF THE BBC. A DISTURBING TRUE STORY.

of people experience these first impressions in everyday life, but in Prison it is a whole new ball-game. The rest of your stay in Prison will usually develop from these first encounters so it is always best to make your mark on any stranger you have been forced to co-habit with from the off. He smiled back innocently at me soon enough so I thought he was safe enough to relax with and we soon got talking together about the usual things prisoners invariably talk about. We got on ok as the days passed and he told me he had put his name down for gym training and was waiting for the gym 'screw' to collect him soon so that he could start his classes. A couple of days went by after this, then one afternoon as he lay sleeping on his bed, the cell door opened and I turned to see who it was. No-one entered so after a few seconds I got up and went to the door to see what was going on. I looked outside the cell and was approached by a gym 'screw' on the landing who asked me if I wanted to go to the gym. I hadn't applied for classes so I knew I would not be allowed to and told this to the instructor. He then told me to make an application in the morning if I wanted classes before attempting to shut the cell door behind me after I had returned to my cell. I quickly realized that my cell-mate was waiting to go to the gym and so I informed the screw of his intentions and asked if I should wake him up so that he might go. The instructor looked at my sleeping cell-mate for a second before telling me not to bother as he was in a rush, as always. The door was locked behind me and I returned to the so-called comfort of my bed. It wasn't such a big deal that my cell-mate had missed the gym on this occasion as he could always go the next time but when he awoke a few seconds later, he did not seem to feel the same way. In fact, he got very irate about it after a few moments of moaning and griping and began to direct his growing anger at

THE DARK SIDE OF THE BBC. A DISTURBING TRUE STORY.

me. I explained what had happened to him while he was asleep in enough detail to reassure even the thickest of arse-holes that it was not my fault he had missed his training, but it seemed the more I tried to explain, the more irate he got. In the end I just thought fuck him! He appeared to be overacting to the situation I realized, probably just to cause a confrontation. Now I was getting irate. I thought about how I had been good to him since we met nearly a week ago and how this sudden outburst of antagonism led me to believe that he was all so false all the time. Now I hated him and wanted to row with him. The party was over. I was thought broadcasting constantly since my incarceration and today was no exception. The whole prison was listening in obviously so now this guy wanted to show off to all the others that he was the man and that I, inflicted with telepathy, was no match for him. A classic case of the big bully, deluded by his own sense of self importance. He began banging on the cell-door and shouting out for the screws to come to his door, like they really would let him out now. He was so obviously trying to intimidate me and soon enough I knew that I was just going to get up and go over to him where he stood at the door and punch his head into a liquid. He wasn't even a brummie. He was from Derby so I wasn't bothered about terminating him now anyway and the more he carried on, the more this was about to become a reality. I was sitting on a chair next to my table by now, having advanced from a lying position on my bed and as I began to rise from the chair in order to carry out the termination of this raving bully, the door suddenly opened and a screw entered the scene. No coincidence. He had been waiting nearby outside the door without interfering in what was taking place so that my mettle could be tested to the maximum. The fact that he realized that my cell-mate was about to

THE DARK SIDE OF THE BBC. A DISTURBING TRUE STORY.

become extinct had prompted him into action. Now he knew that I was a force to be reckoned with and would not be bowing down to any motherfucker in his prison no matter what and so he called my 'friend' outside the cell to speak with him off-side.

I listened partially to what was being said and how he tried to calm my cell-mate down for his own good but I didn't give a fuck now anyway as I was still going to have to do battle with him eventually when he returned just to make sure that that situation would never arise again. Prison is hard, believe me. My cell-mate returned a few moments later and made his way directly to me and began to apologize to me for his actions and expressed regret to me for any ill-feelings I may have towards him because of them. That was enough for me at the time but the damage was done as far as I was concerned and I knew I would not be able to gel with him any more, so I avoided talking to him from then on because I just wanted rid of him basically. Lo and behold, within a few hours of that, another screw appeared at the door and told my cell-mate to pack his things as he was being moved to G-Wing. I was glad to be rid of him and it turned out that every other cell-mate I was two'd up with after that was always about half my size and of no threat to me as I was of no threat to them. It seemed I had been tried, tested and approved. This in itself was nothing new to me and something I had had to maintain for many years now but it soon transpired that forthcoming events would force me to do battle again on three more separate occasions against four different people. For anyone not too sharp with their maths, this meant that one of the battles would be against two different people at the same time. Bad isn't it? Things went a little smoothly after that, except for the trips from my cell to the servery downstairs and back.

THE DARK SIDE OF THE BBC. A DISTURBING TRUE STORY.

It wasn't too bad really, it was just the concern of thought broadcasting amid a crowd of other prisoners going to and from the servery at the same time. After a couple of weeks I knew the opposition was a little weak and so were not a major force that needed to be dealt with as a matter of urgency. I didn't smoke the tobacco in Prison as I hadn't smoked any cigarettes for some time by now and so the word on the landings was that I must be a bit fit and tasty, which all prisoners show respect for. This meant I was safe enough. Of course this was not a good thing for those controlling the conspiracy against me from their studios wherever, in Birmingham and London, but there wasn't a fat lot they could do about it. This initial period of time where there was no opposition to me, made way for some of the cons with a bit of decency in them to drop helpful hints and advice. I knew basically what I was doing and what to expect from others looking at the worse case scenario, but it was always nice to hear it from other people. Especially in Prison, with court-cases pending.

One piece of advice I was given several times was to "trust no-one". I had no intentions of doing that anyway but that was cool to hear. After four weeks in Prison on remand, three of which were spent in Winson Green, I was returned to the Magistrates Court at Birmingham for another appearance to arrange a trial date to be set for each of the four outstanding cases still pending. I appeared in Court 1 again but this time I was up before Judge Morgan, a.k.a the Hanging Judge. He was unpopular to say the least with the criminal fraternity because he was known to always dish out heavy sentences and rarely gave out bail. I had never seen him before but I had heard of him often. On this occasion, he seemed to be of the opinion that the charges against me should be dropped and I was swayed to accept this myself as I was not of the

THE DARK SIDE OF THE BBC. A DISTURBING TRUE STORY.

strong belief that I would expose all the telepathy existing in Birmingham today with drink-drive charges anyway, but after a few anxious moments of realizing that I would only have to go through the whole process again from start to finish with other charges, I decided to speak up and insist that the trials go ahead in any event. Time was slipping away for me and I had to use any opportunity that was available to me. Judge Morgan allowed the cases to go ahead and I was remanded back into custody for one week in order for a trial date to be set. I was back in the Green again and now trying to devise any other method I could to end the telepathy. On the 11th April 2002 I wrote a polite letter to the Evening Mail asking them to end the use of telepathy and then I took it to the Principal Officer on C-Wing and asked for it to be sent by recorded delivery. Alas, it was taken off me as I was told that it was a threatening letter and would be returned to me at the end of my sentence upon discharge. It was never sent. I was to appear in Court again on Monday 15th April which I thought would be for one of my trials and so I also thought I would be getting out or at least bailed that same day. The day before, I decided to order a whole load of tobacco from the canteen so that my cell-mate could collect it for himself as I would not be returning and so neither of us would have to pay for it as he would be impersonating me when he collected it. This is when things started to go horribly wrong. The next day I appeared in Court 1 again but this time I stood before two women Magistrates. Before they entered the Court, the prosecution representative said the word "keys" to me in order to warn me that the keys to my flat were taken. It took me a couple of months to realize what had actually happened. As for this day, the Magistrates agreed to peruse the trials and I was remanded again in custody until the following

THE DARK SIDE OF THE BBC. A DISTURBING TRUE STORY.

Wednesday 24th April when the first two trials would take place in the morning and then afternoon, followed by the next two trials the next day in the morning and afternoon also. I returned to the Prison later that afternoon but the frame-up had already begun. The word was that I had bought loads of tobacco from the canteen but it was kept quiet that it was meant for some-one else. Now it was said that I had tobacco on my hands to deal with and so the telepathy was in control again and everyone was invited to try to rip me off for some tobacco. Tobacco is the currency in Prisons, like phone cards as well these days, it is like gold. I was in reception for a while waiting to go across to the main wings when I was befriended by a couple of con's who were also in reception waiting a return to the main wings as well. One of them who I spoke to was called Osbourne and he named a few people that I knew as well so he seemed to be ok. He told me that he could sell some tobacco for me and I told him that I wanted a £2 phone card for a half ounce packet, which was the going rate at the time. He told me he would sort it out the next day and we left it at that. I was taken across to C-Wing later that evening and spent the night on C-1, the induction landing. I was loaded with food and chocolates as well as the tobacco, so me and my new cell-mate had a good time that night. We agreed to get get padded up together when we went up onto the main landings because I liked him as he did not try to pretend too hard that there was no telepathy going on throughout the Prison. The next morning, my cell door was partially opened by a warden who was asking us both some questions about our induction time, when I soon edged my way past him to take a look at all the activity taking place outside. Lo and behold to my amazement, there was a television camera set-up outside my cell with a camera man standing next to it. I had to laugh at

THE DARK SIDE OF THE BBC. A DISTURBING TRUE STORY.

the situation but I automatically thought that it was the BBC that had landed now that the date had just been set for the trials to take place. I started cursing and swearing at the cameraman in jest mostly just to un-nerve him some. He turned and fled up the stairway nearby and disappeared out of sight before I was quickly ushered back into my cell by the warden and the door closed behind me. A few moments later he returned to my cell door and from outside I heard him declare, "well done, I'm impressed". Some time after that little incident, Osbourne appeared at my door and asked me to pass the tobacco out to him under the door but my cell mate kept warning me not to do it. I genuinely thought that Osbourne would be ok, so I eventually passed out a half ounce packet to him. Then he was gone. Not long after that it seemed like the whole landing started to appear at my door looking for free roll up's. They all wore the same characteristic gleeful smiles on their faces which showed me a piss-take was taking place so I only gave the first two customers a roll up each, two black guys, before I felt uncomfortable enough to see what was going on and told every-one else to fuck off as this cell was not the canteen. My cell mate kept telling not to give nothing away anyway all the time so I began to question my actions concerning Osbourne and it started to appear as if I had been ripped off. People get murdered in Prisons for doing that so I was in a very awkward situation. Then all the voices began in earnest outside my door from the other con's indicating a rip off, but it was when one of the wardens stood outside my door and said the word "DUPED" that the reality really sank in. I felt so stupid.

The whole Prison would be aware of what had happened and were probably laughing at me to boot, so now I had to do something about it. A few minutes

THE DARK SIDE OF THE BBC. A DISTURBING TRUE STORY.

later, the door miraculously opened again and my cell mate was called out to see someone in authority. I told him to leave the door open behind him as he left but the female warden firmly told him not to before she moved off. She must have been concerned for my safety because as my thoughts were always being transferred into the minds of everyone around me, so everyone knew that I was now about to take action against Osbourne for what he had done to me. I had to teach him some respect, but Osbourne is a big lad, much bigger than me, so the general idea was that he was too much for me to deal with. I don't know where people get these silly ideas from.

As it happened, my new cell-mate did leave the door open behind him and so after a couple of seconds I went to it and peeked through the spy hole in order to see if the coast was clear. It was not. There was three wardens milling about nearby and I soon grew impatient because I knew that I had to nip this situation in the bud as I might not get another chance to after this. Then all the wardens turned their backs on me so I advanced out of my cell and up the stairs and onto the main wing. I knew Osbourne was on C-3 and made my way there amid wardens and cons alike who all seemed in a state of mild panic. They knew what was happening but no-one stopped me. When I reached C-3 landing, I saw Osbourne almost immediately. He was employed as a cleaner and was in the process of mopping the floors. He was very nervous looking as I approached him because he knew that it was now time to take responsibility for his actions. He advanced a few steps to meet me before I asked him outright if he had the phone card for me in place of the tobacco I had given him. He mumbled something about he would sort it out later so I asked him what he meant by he'll sort it out later when it was supposed to be done ages ago. Then

THE DARK SIDE OF THE BBC. A DISTURBING TRUE STORY.

he went for it. He suddenly became so over confident that I knew he was performing now in order to mug me off, probably under orders from the 'voices' in his head which emanate from the Evening Mail or whoever. I know some people gain plenty of confidence when they are directed by the telepathy in their head but they don't seem to realize that it does not boost their ability. It does not make them supermen.

But it was when he smirked broadly at me as he started to deny that I had ever given him any tobacco in the first place that really sealed for him. I just punched him full on in the face in front of all his gang and then began to work up a full assault before we both fell on the floor where I still continued to punch his head whilst holding him by the hair. He made no reply worthy of mention as I was too strong, too fast and too vexed for him. Within a matter of 30 seconds, I was landed on by what seemed like two hundred wardens who restrained me to the point where I could only scream my anger at my foe who was led away from me by two wardens as he blurted out the words, "what's the matter with him?" Honestly, as if he didn't know.

I was nicked for assault and appeared before the governor the next morning, 17th April '02. The Governor was Mr. Shann. I pleaded my case of self defense to him and explained how telepathy was operating against me even in his Prison and how the Evening Mail was instigators of harassment directed against me as well. He was concerned and interested in my resistance against the Evening Mail and so now that I thought I would have his support in my plight, I went on to tell him about the television camera that was outside my cell the other day and how it was the BBC that were doing the filming. Bad mistake! As soon as I had mentioned the BBC, it all went pear-shaped as he was a G-Man

THE DARK SIDE OF THE BBC. A DISTURBING TRUE STORY.

and so only interested in saving BBC asses. The guard on duty let out a long sigh and moaned "oh No" as Mr. Shann sprang into action and displayed assertiveness against me rather than the concern from a few seconds ago. I was soon found guilty and sent down to the block for a few days to reassess the situation. In other words I was held in the block to await a hospital doctor who would then commit me to the hospital wing, but I did not know this was going to happen until he landed in my cell down the block the very next day. My word he was quick, normally you wouldn't see a doctor in Prison for a week when you wanted one. I had all my belongings from my previous cell transferred to my new cell down the block and unfortunately for me, one of these was the one thing they all feared most. The most recent written additions for my book, which was now on both sides of about sixteen pages of foolscap paper. I was writing away in my cell when the so-called doctor arrived to see me and he swanned into my cell with such a contemptuous smug grin on his face that I just knew that I was going to the Hospital wing in order to initiate a cover-up. I was correct of course and transferred over on the same day. With less than a week to go before my trials, they had me by the bollox and there was nothing I could do about it. I wasn't bothered.

I still felt it would all come out in court anyway so I bided my time, patrolling the hospital wing on B-Wing like a tiger stalking his prey. They were all prey to me as I knew all things were to turn against me now that I had one foot in the grave, but I was cool as usual and just kept myself to myself as it soon became obviously clear that it was pointless talking to anyone because they would just insult me more or less with verbal under direction from the controlling voices in their heads. These were sick people being used as live bait to antagonize me

THE DARK SIDE OF THE BBC. A DISTURBING TRUE STORY.

and things became a bit intolerable because I couldn't escape them and I couldn't belt them either.. This apart, it was quite sedative on the ward as usual for the first couple of days only after I arrived, but by then, I suppose word must have travelled throughout the universe that I was hospitalized in the Prison and so then, strange people began to appear on the ward with suspicious smiling faces who I presumed to be actors from the Evening Mail. One of them was a black guy dressed up as a chaplain who kept looking over and smiling at me so I decided to ask him straight out if he was from the Mail. His name was Martin Freeman and he smiled again at me when I asked him and said unconvincingly that he used to work for the mail at one time but I knew there was some more to it than that. It was that same day that I phoned my co-writer Michael Miller and asked him to come and visit me so that I could pass out the story so far of the book. I had to get a written form of authorization from the office so that I could pass out my book on a visit, which was done through the correct channels with a matter of some respect shown by the wardens. However, it was a little bit different after word had gone out of my intentions and I arrived in the visiting hall on Saturday 20th April. I was searched first before allowed through and produced my paperwork and the notice authorizing me to pass it out on the visit with Michael. It was a black warden in charge it seemed but he refused point blank to allow me to pass out my book. I argued furiously with him from the off because I knew it was a stitch up but after a few minutes, he relented a bit and told me I could bring the paperwork to my visit with me and let my visitor have a look at it, but I was not to pass it out to him to take away. I agreed with that straight away, but

THE DARK SIDE OF THE BBC. A DISTURBING TRUE STORY.

there was no way my book was coming back off the visit with me. It was going out as planned.

I stayed on the visit with Michael for over an hour with the whole hall full of cons and visitors alike tuned in to my thoughts and our conversation. I wasn't bothered at all really and even played up a bit by deliberately naming the Mail as those involved in the telepathy plaguing Birmingham today. That was confrontational in itself in any capacity but I didn't really care, I just wanted to test for a reaction. All was quiet for a few seconds before a female visitor stood up and proclaimed, "And it's love". That response made me smile as it gave me an insight into the thinking of others, but she was a bit wrong I'm afraid as I did not love anyone who used telepathy as a weapon against me. Not these days.

I did not tell Michael that I was not allowed to pass my book out to him, I just told him to keep it safe under his coat as he left the building. He did this when leaving and I returned back to the secure unit to be searched as we came of our visits. The same black guy was on my case straight away but I think he knew that I had smuggled out my book, they all did. I told him that I had thrown it away in the bin whilst on the visit. He seemed resigned to this and did not want to make a big fuss about it anymore as he probably thought 'fair play' to me, but a white warden who was on duty with him started to perform and asked if he should radio through to the visiting hall and have my visitor stopped and searched. The first screw said "no, don't bother" but the second screw proceeded to radio through anyway. I didn't hear what was said as I had to move on in the queue but it turned out that my £2 phone card was stolen from me for my troubles. Some wardens can be so petty.

THE DARK SIDE OF THE BBC. A DISTURBING TRUE STORY.

I wrote a letter to my friends that week asking for them to stand up in Court for me as a witness to the existence and misuse of telepathy that was in use against me. I had already given my previous solicitor Brian Pugh four written statements to that effect but I really needed a show of support from living witnesses in Court. I wrote to Des Mulcahy, Peter Flynn and Michael Miller.

I had no time for a reply from them to say whether or not they would be there at court for me so when I was taken back there for trial no. 1 on Wednesday morning 24th April, I was full of hope and expectancy. I stood in the dock before the presiding magistrate who was a Mr. James and I felt a whole lot better when he informed me that my three physical witnesses had turned up in Court for my defense. I pleaded my case that I was driving my vehicle under duress from the voices in my head emanating from the BBC and so I was not guilty of willingly driving my car whilst over the limit. An authentic and legitimate defense demanding a not guilty verdict but not to be, not with an expose of the BBC involved as a result to boot. In this case, the punishment did not outweigh the crime it appeared. My three witnesses took it in turn to say their piece about the telepathy in Birmingham and the BBC were even named as instigators before they were allowed to sit at the back of the Court when they had finished so that they could listen in on the rest of the case and its outcome. I mentioned the involvement of the Evening Mail myself as cat calls and laughter followed me from underneath by the stupid actors trying to disrupt me, but I just snarled my contempt at them as they had no effect on me with all the knowledge that I knew and was ready to dispel. My mission in life was still to expose the BBC since it began years ago and anyone or anything else

THE DARK SIDE OF THE BBC. A DISTURBING TRUE STORY.

was secondary. The Evening Mail was now a very close second though. The plot thickens.

At the end of the trial during his summing up so to speak, Mr. James admitted that it was probably true that the BBC had left someone in charge to take over where they had left off in Birmingham all those years ago, before saying that he would take a few minutes to himself to come to a decision. It was all bullshit really. He left me standing in the dock by myself so that the whole world and his uncle could listen in to my thoughts for a while before he would announce his verdict. His guilty verdict at that.

When he was ready, Mr. James looked up from his studious posture and informed me that he was ready to come to a decision. I looked at him straight in the eyes as he began and a huge, smug grin appeared across his face as he could see himself on camera before the whole town in his mind's eye but when I heard him come to the part where he said he finds me guilty as charged, I got so fucking angry that my teeth clenched hard together like I had lock- jaw or something and I dropped my eyes away from him to the ground as I knew he was smirking to himself for all to see around the town. I cut him off short from his moment of glory just to piss him off. No sooner had I done this than his voice changed from a tone of confidence to that of a nervous, shaky drone that belied his realization that I had knowingly cut him off camera.

I was remanded for 21 days to reappear for sentance to be passed and taken away in handcuffs. I was still so very angry as I was led off that I shouted out to my co-writer Michael Miller not to worry about these bastards as we would still write the book. He smiled back at me as though he thought I was still mad

THE DARK SIDE OF THE BBC. A DISTURBING TRUE STORY.

enough to carry on but the thought of giving up had never, ever entered my mind.

There was no second trial that afternoon as the first trial was the important one that would set the stage for the rest of them and that had run hours into overtime leaving insufficient time to start the second. I re-appeared back at the same Courts the next day but this time it was the trial concerning the involvement of the Police in the conspiracy, where I was arrested for allegedly being drunk in charge of a motor vehicle (not driving), after the day out that begun with the Spanish properties exhibition in Merry Hill, outside Birmingham. Des and Pete were both there again to go as witnesses but I already had that feeling of deja-vu and knew that it was going nowhere so I just told them all to forget it as I would be pleading guilty to all the rest of the charges. I really could not be bothered to go through all that farce again for nothing as three Police stood nearby trying their best to look intimidating and menacing. I pleaded guilty and was remanded again for 4 weeks this time for sentance to be passed. I returned to the hospital wing of Winson Green Prison for the duration, seemingly a failure in everyone's eyes, but to me it was just another anticipated hiccup in my mission to free myself from television tyranny. The whole of the hospital wing started to fill out to full capacity with other prisoners hell bent on giving me a hard time as actors and voices for the perpetrators of telepathy against me, including actors for the Evening Mail and they operated all day long around me to such an extent that I couldn't talk to anyone at all as the whole wing was there for use against me and me alone and when they were all locked up in their cells at the end of the day, the hospital screws and the cleaners would begin their tirades against me and even after they were

THE DARK SIDE OF THE BBC. A DISTURBING TRUE STORY.

finished, the televisions were left on all night long outside my cell to harass me from the studios, so my brain was under constant attack for 24hours a day, seven days a week. It was all poisonous remarks made to demoralize me and to break my heart so that I would never try to expose the dark-side of television again. They were really out to do a number on me. This was to persist now until I left the Prison and not to stop until I was 'pacified' as they called it and forced to admit that I was just a wanker. Pleasuring myself in that way, the way everybody else does was a thing of the past for me by now obviously, though I was normally used to the charms of a pretty woman for the most and I definitely was not a wanker in the colloquial term of the word but this is how the gutter press from the evening mail went about describing me I suppose, for the simple reason that it was them that bottled out of a showdown with the BBC, not me, as described in the earlier chapters of this book. They were using me as a scapegoat to cover for their own inabilities.

By Saturday 4th May, I was so pissed off with what I was being subjected to, that when some dick-head appeared in front of me and declared to everyone "we have landed" (as if he was the leader of an army against me), I just got so irate that I really wanted to just punch his face in. Everyone else were a little less confrontative than that but now that I was really thinking about doing him in as well, he was ordered to appear in front of me again by the voices in his head in order to see how far I would go. I could tell this by the way he turned up out of the blue pretending to be using the phone but his eyes were fixed on me all the time as if he was daring me to do it. It didn't take long for me to go through with it as I was vexed enough anyway, but as I advanced on him looking obviously like I was going to belt him, he did not take his eyes

THE DARK SIDE OF THE BBC. A DISTURBING TRUE STORY.

off me even though his face was tightened and contorted in anticipation of a punch in the face. He was provoking me right up to the very last second. It must have been like playing 'chicken' to him, much the same as the way people behave using words and intimidation on the streets. He was cringing in anticipation by the time I let off a big right hook that connected squarely on his jaw and nearly sent him off to dream-land as he landed on his arse. I immediately jumped on top of him to continue the assault as I had already hit over-drive but the wardens were all over me within seconds and I was hauled rather hastily away to the block for the second time. This time I wasn't even nicked at all but returned to the hospital wing by order of the female Governor who came to see me, probably so that I could face the consequences of my actions. And face them I did.

One of the lads working on the servery on the hospital wing was called Mark Porter. He was a big lad with a shaved head the same as myself and although I didn't recognize him at first (not until some months later), he was the very same lad who I saw approach me when the BBC film crew were outside the hospital I was in back in 1995, Small Heath Health Centre. He had a work-mate called Simon Orton (I don't know the correct spelling), and they both turned out to be friends of my nephew, Shaun. As they were both on the servery and had a lot of freedom of movement, they were both well involved in the telepathically controlled actions against me and we nearly came to a head on one occasion, but as they were some of the few people that could have a laugh as it were about what was going on, nothing very serious took place. In fact, not long after I was returned to the hospital wing after the last fight I was involved in, I got myself into another fight on the exercise yard. This time it was against two

THE DARK SIDE OF THE BBC. A DISTURBING TRUE STORY.

big fella's. One of them was a tall, light skinned Nigerian called Andrew Attah and the other was some Asian body builder whose name I never came across. I was walking around the exercise yard on my own when Attah stood in front of me and yelled at me in a very nasty way to get out of his way. I think the object of the exercise was to 'bottle' me (frighten me), which would then in turn leave me humiliated and so open to all manner of nastiness from the rest of the prisoners. As it happened, I was more alarmed than frightened for about half a second, then I just steamed into Attah who was directly in front of me. The Asian guy was standing about a yard behind Attah. We ended up belting hell out of each other in a clinch, him with his left arm around my neck and me with my left arm around his neck. The Asian guy was also steaming into me as well but I was so angry and fired up in a split second that I didn't even feel anything. I got so mad because I couldn't get free that I actually 'saw red' as the old expression goes. I just wanted to bite his ear off in my anger and I could feel my open jaws inching towards his left ear ready to bite. The next thing I knew I was being pulled away from the fracas by Mark Porter who was telling me "no Tony no! You mustn't fight Tony!" The wardens just stood around watching it all happen without remotely trying to stop it, like in one of those corrupt American jail movies. When they did intervene, I was taken to my cell to cool off so to speak but again I was not nicked for it so the matter was not brought to the attention of a Governor and so, it was not officially recorded. Wouldn't you know.

I was still waiting to be sentanced at Birmingham Magistrates Court but in the meantime I got fined £735 at the same court in my absence for other more trivial motoring offences. I didn't even know about these matters until a

THE DARK SIDE OF THE BBC. A DISTURBING TRUE STORY.

long time afterwards. When I finally did appear at court on 23/05/02, I was now to be represented By Simon Bailey who was the top man in Carvers Solicitors. I spoke to him in an interview room underneath the court-house and with all the sexual slurs directed at me still fresh in the air, he just wanted to get all matters dealt with by way of "getting all this crap dropped" as he put it. I agreed with him at that time even though most people were urging me to appeal against conviction and sort out the media that way. I'd had enough of relying on petty crimes to do an expose so big so I left it up to Mr. Bailey. When I eventually stood before the presiding magistrate, he also allowed the cover-up to continue by sentancing me to a hospital order. Whilst in prison, the officials had tried so hard to make me take medication of my own free will so that it would look as if I was weak and ill and needed medication but I was adamant not to do so. I never got any respect for the shit I went through. Now I was to be returned to the prison hospital and to all that shit again for a maximum of thirty days until a proper hospital bed was ready for me. Any-one else would probably get seven days. After I was returned to the prison, I just could not be bothered to fight off the conspiracy after all the sleep deprivation and everything I went through including the finger up the bum routine almost daily, so I just used a bit of common sense and let them think they had won their silly little battle against me in order to get out of there. It worked a treat and so on 31/05/02 I was transferred to the new Meadowcroft Hospital in Winson green after only a further seven days more prison, instead of thirty. I was glad to see the back of that place but the same shit awaited me in Meadowcroft Hospital just around the corner. One of the dick-heads there supposedly from the Evening Mail as I was told by a nurse, was called Paul Meadon. He was one

THE DARK SIDE OF THE BBC. A DISTURBING TRUE STORY.

piss-taking bastard from a distance but at times I found him to be quite funny. The only thing was, he couldn't take as much as he gave so after about a week, I decided to have the battle with him the next morning. When I got up the next morning, he was gone. Lucky him!

Not long after that little episode, I spoke with one of the patients there also called Paul about his problems with his mental health and he mentioned in passing how he had taken his case to the European Court Of Human Rights in Strasbourg, complaining of torture and the like. It took a few seconds to register, then it hit me like a hammer. What the hell was I doing trying to do an expose of Birmingham media in a telepathically media controlled Birmingham, when I could take my case to an unbiased and proper court dealing with human rights issues in Strasbourg, France. I was in a bit of an excited fluster for a second or two but then, within minutes, I got Paul to write down the address and phone number of the court in France and he also directed me in how to approach the issue. I gleefully sent off a letter to the courts detailing my complaint against the BBC and the Evening Mail Newspaper that very same day but I arranged a visit from my co-writer and friend Michael Miller and got him to post it from outside as I could not even trust the staff to allow it to be posted from the inside. My mail to my Mom had 'gone missing' before as they kept me segregated from my close ones. I gave my home address to the courts to reply to and bore the brunt of my shitty existence at that time with anxious anticipation of a response from dear old Francais. I had a light at the end of the tunnel once again. By 11/05/02, I was so angry and pissed off with all the telepathy going on and with all the patients and staff on my case that I decided on a little holiday from

THE DARK SIDE OF THE BBC. A DISTURBING TRUE STORY.

it all. I went to my room at about 10 o'clock in the evening and took the chair that was there and proceeded to smash my way out of the window in my room. I really just needed some space to myself for a while away from continuous lock-up to make a phone call or maybe get laid but it turned out that the window was fucking unbreakable. After the thunderous bang of the second attempt, I quickly realized the futility of my ways and jumped back to normality (so to speak) into my bedroom. I knew that I was still thought and sight broadcasting and that every-one knew what I had just done so just for the craic, I stuck my head out of the bedroom door into the corridor and exclaimed aloud to them all down the other end of the corridor "it wasn't me!" Then I jumped back into my bedroom where I pissed myself laughing for a few good minutes. Word of this must have got back to my social worker at that time Mick O'Hanlon because he appeared at the hospital the very next day with a big smile on his face as if he knew what had been going on and told me I was being moved out to an open ward back in Small Heath, where I had seen the BBC on my last detention there in 1995. I felt happy about that as it was one step from going home.

I decided to go the whole hog in my fight to expose the illegal use of 'telepathy' against me and asked if I was allowed to have my computer brought in. (I had a brand new computer by this time). I was determined to finish my book and also, I was determined to continue my peaceful campaign as well by starting up a petition of people's signatures of those against telepathy. Now things were starting to look up again as I trained every day in order to maintain my physical prowess in preparation of any battles as well. On 14/06/02 I was finally transferred to Small Heath In Patients back in Small

THE DARK SIDE OF THE BBC. A DISTURBING TRUE STORY.

Heath. Four of the nurses in charge there were called Hilary and Suzanne and Patrick and Colin. They seemed to like me because I was upfront and carefree and helped me at times amid the huge conspiracy in and around me even though they were Government men (G-Men as I fondly used to call them). I liked them too because they were fair with me and I was soon allowed to have my computer in hospital with me. I immediately started back on my book but unfortunately, I was caught drinking a can of lager on the 20/06/02 and ended up being sent back to the dreaded Meadowcroft on 21st. My mate Peter Flynn visited me there within a couple of days and his name was also added to my petition. On 8th July, Michael Miller visited me again after I asked him on the phone to collect my mail from my flat and he brought tidings of great joy. There was a large letter from Strasbourg! In it was an official application form to be filled in and returned to the courts. It was a bit complex so I left it for a while in order to seek professional advice. I also got Michael to sign my petition. I had been hospitalized for a total of about two years on my last two detentions but this time I had my book on computer, my petition going strongly and my case evolving at the Court Of Human Rights in Strasbourg. The G-Men did not want to get involved with my illegal detention now and even Dr. Kennedy sorted me out by allowing me to go back to the open ward in Small Heath again. I was returned there on 10/07/02. I met a girl called Gemma who was in the hostel next door and it was lust at first sight. I chatted her up for a few days using my charisma and charm and after about a week, we arranged to go to my flat for some sexy bits. I hadn't had sex for about five months so I was well up for it, as was she. I even bought her some white stockings and suspenders etc from Woolies across the road for the occasion on the next day but when play time

THE DARK SIDE OF THE BBC. A DISTURBING TRUE STORY.

came around, she had been discouraged from it by the stinking media forces. It was sad but her loss as she was then eventually turned against me and tried to wind me up by going with other men. She wasn't that pretty to pine over. I persevered in my plight to maintain a normal existence and was eventually allowed to have weeks and weeks of leave from hospital by my then consultant psychiatrist, Dr. Siva. He had been in charge of my care for about two years when I was outside and he seemed to like me as well. Word had gone around that I was writing a book about my exploits and I had to pretend to the doctors in charge that it was just fictional in order that they could not prove I had a mental illness by saying my writings were proof of my so-called delusions. I had to keep up this pretence so that I would soon be discharged but one day I had an interview with Dr. Siva and other nurses during his ward round. He asked me outright about my book and I began to prattle on with my bull-shit of how it was all fictional when he suddenly interjected with the word "fact!" Then he started laughing almost uncontrollably at the situation and caused me to laugh aloud as well because he did not give a damn. He was really funny in his ways. He knew he had nothing to hide as he had never done me any wrong and seemed proud of me for using my intelligence which he also seemed to enjoy. He knew I was 'outing' one serious media wrong-doing. Time went by quickly after that and I had the chance to sort out my bills at last and my finances. I used the phone in the office to see if I was still getting my benefits paid as the benefits office did not know I had been incarcerated. Not officially anyway. As I was talking on the phone about my benefits, it seemed like a hundred people rushed into the office and started milling about suspiciously. They were the dreaded Evening Mail and they were out to fuck me up with my benefits. They

THE DARK SIDE OF THE BBC. A DISTURBING TRUE STORY.

waited until I had finished on the phone and then they all disappeared again as quickly as they had arrived. It turned out that they informed the benefits office of my detention in prison and hospital and got my money stopped out of sheer nastiness. I was a bit upset by that. Anyway, I was soon allowed continuous leave from hospital by Dr. Siva after that so my benefits were soon restored to me as I was no longer classed as an in-patient. I returned to the hospital one day to collect the medication I was forced to take and I went into the office to let them know I had arrived. I stood talking to a member of staff in there for a couple of minutes as I waited, when my eyes caught sight of an official memo that hung from a shelf at eye level. It was dated 16th August 2002 and it read in brief that members of staff were reminded that they were not allowed to talk with the media about any of the patients that were in hospital care and that patients information was confidential etc. It was signed by Hilary. It was a bit late to protect me from the media snoops but at least it gave me surety of who my main antagonists were.

I was officially discharged from hospital on 16/09/02.

BREIF FREEDOM.

THE DARK SIDE OF THE BBC. A DISTURBING TRUE STORY.

I was on permanent leave from hospital about six weeks before I was officially discharged and in that time I met an old acquaintance called John Howell. I had known him for about a year on and off and he was in Winson Green Prison at the same time I was some months ago. He told me he was homeless again after just coming out of prison so I let him stop at my flat for a while until he got himself sorted out as a favor. He promised to look after my flat and help finish off the decorating for me but alas, it was all bull-shit. He never did a spot of cleaning, not even his own mess. Before I got pissed off with his ways, he introduced me to a friend of his at my flat, called Jane Henry. She was a half-caste girl in her early thirties and she was a working girl now operating in Small Heath. She took a shine to me straight away but as I had never had sex with a working girl before, I was a bit cautious. However, as I had also not had sex at all for about five months, I took her into my bedroom when John went out to the shop for my first blow job but got her to finish off with a hand job in the end as I was near to exploding. Hee! Hee! She did it because she fancied me, not because I paid her. She turned out to be near enough homeless so I let her stay at mine a few times as well. I was a bit cautious as to whether it was all a big set-up but even though they both were making the odd comments into my microphones, I let them stay because they were not too bad. Eventually, I became tired of John's false promises and decided to throw him out from whence he came the following day. I was thinking this to myself in the living room as John lay on the sofa one night and I heard him moan to himself "oh fuck". He had taken the piss for long enough now and the voices in his head told him he was getting evicted tomorrow. I went to bed feeling a little more happier that he would be going the next morning and fell into a deep sleep. He

THE DARK SIDE OF THE BBC. A DISTURBING TRUE STORY.

awoke me the next morning with his movements about the flat so I got up in time to meet him as he was leaving. He told me he was just going down to his Mom's flat for breakfast and would be back within the hour so I decided to break the bad news to him when he returned. I was in no mad rush to get rid of him as I still felt a bit sorry for him but I was a bit foolish. I went into the toilet for a wee as he was leaving and didn't think he would do what he did next. He had one of my jackets on his back as he was leaving but when I was in the toilet, he reached into the cupboard in the hall-way by the front door and stole my leather jacket as he left. I cannot say that he was operating under instructions from the voices in his head but I would lay money that he was. That was the last I saw of him for a long time but as there was now a catalogue of sneaky crimes being committed against me due to the telepathy, I decided to start reporting them all to the Police, even though I knew they would not do anything about them. I was still laying the groundwork for a case in the courts and these crimes on record would help me eventually.

Then there was my next door neighbor, Lincoln. He had lived next door to me for about two or three years and was ok but he was controlled by telepathy much the same as every-one else, except that he was used to annoy me at all times of the day and night by making comments about my thoughts and also by coughing outside my flat. I kept meaning to talk to him about it but felt a bit silly really as he was always nice to me whenever we met. He had a girlfriend staying with him most of the time and she was just the same. It was getting to a point where I really had to say something because sometimes I just felt like running outside to him or her and screaming at them. Most of the time

THE DARK SIDE OF THE BBC. A DISTURBING TRUE STORY.

they were used to do it when I was in bed so you can imagine how upset I used to get.

After John had gone, I let Jane stay at my flat almost all the time and eventually I let her sleep in my bed but we never had sex. She was good to me in her own little way and did enough of her fair share of the cleaning and even gave me the odd £20 when she had it to give. I grew fond of her eventually as a friend and even had a proper blow job by her on the one occasion in bed. It was good and soon enough I was wanting full sex with her even though she does not usually do full sex. I knew I could charm her and as I lay in bed one morning after she had gone out, I was thinking about it to myself, not caring if I was still thought broadcasting, when I heard a young lad outside say "he makes love". So I was broadcasting and my little friend's comment was also broadcasted as well which I discovered when I went down to the shops a little while after that. Every-one was being nice to me because they had heard I was a lover, and when I passed Lincoln on my return to the flat, I heard him say "you're all right Tony". It seemed every-one was awaiting the day I would have sex with her and God only knows the perverted stories they were fed by the 'dark side of television' in the interim. Unfortunately, Lincoln and I were to fall out and do battle not long after I had sex with Jane.

It happened on a Saturday night after I had been drinking indoors with Jane and another two lads, one called Paul, alias 'Beaky' and the other was called Barry Irvine. I had known Barry for many years. I would not permit myself to drink outside my flat anymore after my previous arrests for drink driving because I now knew that the instigators of the telepathy at this time, the BBC , could black me out of consciousness at the touch of a button, so I

THE DARK SIDE OF THE BBC. A DISTURBING TRUE STORY.

was wary of the situation to say the least. We three left my flat to go down to the local shop in order to get some more cans, when upon our arrival we saw a man there supposedly from The Evening Mail who was looking well stern. I didn't care if they hated me or not at that time because I was all 'loved up' from the drink and I proceeded to lift him up in the air with merriment. Then I was blacked out by the studio merchants but nothing untoward happened as they must have hoped would happen and I was eventually released from blackness after I had opened the door to enter my flat some time later. I then noticed Lincoln at the bottom of the stairs leading to both our flats and I told him that we would have to have a talk. He must have known I was going to confront him about his involvement against me but as he had not ever been nasty to me, I was indeed only going to talk to him and ask him not to be used in that way. He saw it differently though. He must have been groomed to do battle in anticipation of this day because he just smiled nervously up at me and said, "I'll talk to you later!" Then he disappeared into the night. Then I was immediately blacked out again for the duration of the evening but found out later that all went well with us boys and Jane and we all had a good time that evening. I added more signatures to my ever growing petition.

I awoke in bed from deep sleep at four o'clock in the morning. I had been blacked out continuously until then and I was as groggy as hell when I awoke alone without Jane by my side. I stumbled around the bedroom stark naked looking for my slippers when I heard a loud knocking on the door. I immediately thought it was Jane so I didn't think twice about answering the door in the noddy. It was not Jane, it was Lincoln and he had come to 'talk'. He had also come armed with a blade of some kind and before I knew what was

THE DARK SIDE OF THE BBC. A DISTURBING TRUE STORY.

happening, we were both brawling on the landing outside my flat. Lincoln is not a very big lad compared to me but he is known as a 'nutter' around our area. I do not need weapons to back me up. I was so very groggy at that time in the morning but I still managed to throw him up and down the wall outside our flats. I was still pretty powerful even in that groggy bare-assed state. One time I lifted him up above my head from under his arm-pits but he slipped out of his jacket and onto the floor where he proceeded to stab me in my right leg and right foot. When he backed off, I saw him struggle for breath after all that commotion as he stood near the stairs and I remember thinking God, he hasn't got much puff but we were probably wrestling away for a good few minutes. I don't think I threw any punches up until then, I was obviously not feeling active enough at the time, but when he backed off and I suddenly fully realized I had been in an important fight and my opponent was not lying decimated on the floor, I was just about to start throwing punches at him even though I knew now that he was armed, when he backed away and went back quickly into his flat opposite me. My interest in prolonging the battle went away as quickly as it had arrived and I returned to my flat as well. I think Lincoln must have been peeved with his performance because he came out and started yelling and banging on my door. I knew he was armed and I started to get mad with him and shouted to him to "fuck off now, or I will stick a knife in you!" He got the message straight away and returned to his own flat. I returned to bed with blood on my feet and slept through until Jane woke me at about nine o'clock in the morning. We were in the living room with the TV on and I was telling her what had happened, how Lincoln had attacked me at four o'clock in the morning when I was starkers and it seemed amusing to me then but I said to

THE DARK SIDE OF THE BBC. A DISTURBING TRUE STORY.

her that we would probably have to have a 'straightener' without weapons when a man on the TV with a big smile on his face said, "OK, we'll make arrangements for that".

Television people took over the situation and Lincoln was indoctrined into the studios (I still thought these broadcasters were studio based at this time) where he was allowed to use all the technology of the 'Dark Side Of Television' against me in an effort to break me in the usual bull-shit way. They would never break a fighter's spirit.   I heard his voice amongst others projected into the huge electrical field that encompassed and surrounded my brain and his piss-taking and abuse was intense at the beginning but as time went by and he realized he would not break me and would soon have to face me in real-life again, he seemed to acknowledge defeat. He admitted to all that I had won one time when I was out shopping at the Swan Markets in Yardley and then I saw Kathy Hooper (little Josh's mother who I had known for years) make an appearance and say into my microphones "see". She was saying that now I should be able to see my opponent as I was the stronger and would bash my opponent in a real fight. I was losing interest in a 'straightener' by this time because I had been on 'red alert' for some four weeks by now. I knew I could beat Lincoln in a toe to toe anyway as I had already done so really just weeks ago.

On 16/09/02, I filled in my application form for the Courts Of Human Rights and posted it off straight away. As I was filling it in, I heard Lincoln's voice projected outside my window and he seemed to be enjoying the situation. I heard him say aloud with joy in his voice "and he's writing a book". He wasn't all bad, just foolish for getting involved that way with the illegal activities of television. He started to make 'appearances' soon after that.

THE DARK SIDE OF THE BBC. A DISTURBING TRUE STORY.

Outside my window to begin with, usually when I was undressed in my room or sometimes just after I had just had a serious workout on my punch-bag and was not going to be on the offensive for a good while. So I was on and off 'red alert' even after he had left the studios of the BBC and these appearances were called 'heads', whereby we would encounter each other head to head in real life. It was slowly annoying because every-one expected me to fight him even at times when I was busy or I wasn't in the mood. I would later learn that these people were not working from the BBC studios but that they had this illegal broadcasting equipment installed in their homes. That is how they could stay tuned into me 24hours a day.

    I telephoned the Courts in Strasbourg to make sure they had received my application form, even though I had sent it by recorded delivery (I was taking no chances), and I spoke to a French woman on the phone. She acted as if she was expecting me and knew who I was so I grew a little suspicious. She told me to phone back in half an hour so I waited and did just that. I got a recorded message from the other end this time saying that this number was not available at this time and I should phone again later. I grew even more suspicious. I phoned back again after about an hour and this time I spoke to the same woman who put me through to another department. A man answered this time and he was so obviously English with his accent that I was even more suspicious. I told him about how I had been put onto an answering machine when I had been told to phone back in half an hour etc and he seemed to be taking the piss when he said "that's strange isn't it?" He was then a bit curt and told me that if I had not received a reply to my application form by three weeks, then I should phone back. I left it at that but I was very concerned as to

THE DARK SIDE OF THE BBC. A DISTURBING TRUE STORY.

if he was from the Mail or the BBC and was over there to intercept my application. I was so concerned that I phoned back again some days later to talk to some-one else. I spoke with a Mrs. Armianova eventually and she was very helpful it seemed and told me my application would have been received in another department, section 9 and that I should phone back and speak to them. I put down the phone in order to do just that before realizing that I only had the one number for the courts. I phoned back anyway in order to seek directions to section 9 and guess who should answer the phone? The same English speaking man from a few days ago. I told him I had just spoken to Mrs. Armianova and was trying to get through to section 9 in order to trace my application form. He immediately told me that "this is section 9" . I was being given the run around because it had all gone telepathic over there in France. I had phoned this same number about four times by now as my phone bill proves and it was not a direct line to section 9. Also, every time he talked, he spoke very quickly and then stopped dead as if he was rushing so that he could stop and listen to my response, which gave me the impression that he was putting the onus onto me to do all the talking because everyone was listening to our conversation. I was clever enough to make him tell me his name. He said his name was Stephen Phillips but he could have been clever enough to give me a false name I thought. However, he did start to get snotty with me and threatened to put the phone down on me which was the final nail in his coffin as a legal worker for The European Courts of Human Rights would not dream of doing such a thing, so I screamed down the phone at him "PISS OFF THEN!"

This conversation took place on 17/10/02 and I immediately got on my computer and went to the courts own website. I downloaded and printed off a

THE DARK SIDE OF THE BBC. A DISTURBING TRUE STORY.

new application form, much to the horror and respect of the Evening Mail agents I encountered on the streets soon after, then when I returned to my flat I worked through the night filling it in ready to post it off the next day, Friday 18th October and it was posted off without any confrontations. I 'ruled' as the media would say.

The following week, I was in Birmingham Magistrates Court again to answer to the unpaid fines I had that were imposed in my absence while I was in prison on remand some months ago. The whole court-house was tuned into my thoughts and my vision as it was said that I was to expose the illegal activities of the BBC et al with this case that was to be re-opened. I eventually got the services of the duty solicitor that day, Ian Hampton and we went in to court with me only intending to get my outstanding fines quashed as it was said that I was suffering from a mental illness and should not have been fined in the first place. I let them say what they wanted to as it was in my interests to get the fines squashed. The presiding Magistrate greeted my new solicitor and said "so you're taking over where Brian Pugh left off?". The case was adjourned in order for Mr. Hampton to acquire the necessary papers from Mr. Pugh and I went off home feeling content that I had a new champion on the case. Lincoln did not appear at court to 'head' me off as it was said he would, so I knew he was not involved anymore as a threat to me and so I treated our little miss-adventure as over. I had more important things to do now but I still kept myself toned-up just in case. I did see him as I was going into my flat the following evening but we just shouted a little at each other before going into our separate homes.

THE DARK SIDE OF THE BBC. A DISTURBING TRUE STORY.

As I was working on my book around this time, I noticed that whenever I wrote, or even thought about writing anything to do with the main controllers of the conspiracy around me, my computer would suddenly freeze and I would have to wait patiently until it functioned again. This taught me that my antagonist's had the ability to interfere with my computer at their leisure, though at first I thought they were accessing it via the mains electricity supply through the plug socket. I was quite wrong in my assumption. As I continued to near the end of my story, my antagonist's really started to show their capabilities within the dark side of television.

First of all, they would block certain commands I gave to my computer, such as 'save' and 'open' certain documents, then they would start to cause my lights to go out in my living room, then come back on again when I either laughed aloud at them or commended them on their abilities. They also controlled my video recorder so that I could not play or record when I wanted to, not to mention my TV or my stereo equipment. With a little intelligence, I deduced that they could manipulate any electrical equipment around me not via the mains sockets, but by enlarging the electrical field attached to my brain to include the said equipment, then it was as easy as operating a remote control for them. I also learned that they could not cause any electrical equipment around me to function or operate at their command, they could only 'block' the functions or commands that I myself gave any electrical equipment. Fascinating huh?

I also realized by now that because the people that were continually monitoring my thoughts and actions were present in my head almost 24/7, they could not be operating just from the studios based at the BBC studios or at the offices of the Evening Mail, they must have had broadcasting equipment installed in their

THE DARK SIDE OF THE BBC. A DISTURBING TRUE STORY.

own homes like the ones I saw on television once that were only the size of a small phone box. These people did not seem to eat, sleep or move away from their equipment so it can only have been accessible to them in their own homes. Now that is what I would call 'spreading like a cancer'.

I had a new court-case pending in Birmingham as well again now for the outstanding fines, but three days later I was sectioned again under the mental health act and whisked off back to meadowcroft hospital in Winson Green. They came for me at my flat on Friday morning 25th October and they were armed with an unsigned warrant and a battering ram. I had only seen the medical services the day before and three days before that and everything was ok, but I did get the impression that Dr. Karan and my social worker Linda Bedford were trying to 'put me away' when I saw them the day before. There was a loud knock on my door on the Friday and I looked through the spy-hole to see Linda with another doctor in waiting and they were accompanied by about eight Police officers, three in uniform and about five C.I.D in plain clothes. I told them through the door to go away and come back tomorrow as I had some-one in (I was lying), because I knew straight away that they were going to section me. They did not go away and continued knocking for a few more minutes before the Police sprang into action and battered my door down with the battering ram. They all surrounded me in my flat like I was some dangerous psychopath or something and when I asked what right they had to smash my door down, one of the uniformed PC's flashed the warrant at me but wouldn't let me read it because it was unsigned. He was PC 7856 Jarratt. The doctor asked me questions about the BBC and telepathy etc and I told him it was all an illness I had years ago but I had been 'well' for a long time now.

THE DARK SIDE OF THE BBC. A DISTURBING TRUE STORY.

They blatantly talked conspiracy talk in front of me after that with the Doctor, Dr. G. Hanna and Linda being on Birmingham's side and the Police being on the BBC's side. I was sectioned anyway and then the warrant to search for and remove patient was signed in front of me. The best thing was that the false reason the warrant was issued was because it said that I was unable to take care of myself and was living alone at my address. Unbelievable nerve or what? I was a bit foolish really because I heard the word "paperwork" on the TV the night before and it was telling me to produce my paperwork I had amassed for my eventual court case to these blatant kidnappers. That's basically what they were.

For my eventual Court-case, I had five signed witness statements, a petition with eighteen signatures on it at that time and my book now on compact disc, amongst a truck load of incriminating paperwork. I should have shown it to the doctor there and then in front of the Police and it would have been literally criminal of them to section me after that. As it happened I was still trying to keep all my evidence secret until the court case but I did not let them take me away without taking all my evidence with me and disarming my computer so that they couldn't get at the hard-drive. I was a bit clever then after all. I was whisked away to meadowcroft where I would stay for only three weeks this time.

THE DARK SIDE OF THE BBC. A DISTURBING TRUE STORY.

HOSPITALISED AGAIN.

My main concern was that my front door had been smashed in and whether it had been properly secured in my absence. I had to phone my dad and ask him to pop round and have a look at the damage and see if it had been repaired

THE DARK SIDE OF THE BBC. A DISTURBING TRUE STORY.

with a new door or if it had been boarded-up properly. He reported back some hours later that the door had been boarded-up with chip-board and was not very secure at all really. Being burgled and having my computer stolen was my main worry but luckily enough it did not happen. A new door was put on in my absence after about two weeks. I met a fellow in-mate called Carlton Davy during my short stay back in Meadowcroft on this occasion and he was one of only a few in-mates that was ok to confide in as he was quite sane like me and eventually he also signed my petition against the big, bad BBC. I was also worried that a reply to my application forms sent to the Court of Human rights in Strasbourg recently would be 'returned to sender' as they would not have been able to have been delivered to my flat and I would lose contact with my court-case until I was released from my incarceration. I worried about that for the three weeks I was in Meadowcroft until I was transferred back to the open ward in Small Heath In-Patients. Nothing of interest or concern happened in Meadowcroft on this occasion, my nearly completed book on compact disc was a major concern now.

I was transferred to Small Heath In-Patients on 15th November 2002 and quickly settled into the regime once again. I was not allowed to leave the unit until I had seen the consultant psychiatrist who would assess me first to make sure I was fit enough to be allowed out. I saw Dr. Khan on Monday 18th November and he immediately gave me two hours leave from hospital per day. This was upped to four hours per day the following week. In the meantime, after some concern for my mail being delivered to my home address, especially mail from Strasbourg, I went back to my flat alone and collected all the mail that lay there on the mat beneath my letterbox. There was a big one from

THE DARK SIDE OF THE BBC. A DISTURBING TRUE STORY.

Strasbourg lying there smiling up at me and I knew then that all was going well. The letter read that the courts had received both my applications and my case continued to progress. I was overjoyed.

The letter was addressed from Clare Ovey, the legal secretary assigned to my case and I read it over and over to make sure I was not dreaming. I wrote back to her personally on two separate occasions soon after that with more details of the high-ranking conspiracy against me and she replied to both my letters to tell me that they had been received and were now on file with the rest of my case-papers in readiness for my eventual hearing at the Court Of Human Rights in Strasbourg. My day of reckoning was soon to establish itself at last.

My appeal against section under the mental health act was also acknowledged at about this time and the papers I received from the appeal tribunal authority told me that my appeal was to be heard on Friday 6th December 2002. My hopes of release from incarceration lay on that day. I acquired the services of a solicitor from Ashby De-La-Zouch, just outside Birmingham to represent me as my faithful mental health solicitor Jon Lloyd, was no longer practicing. My new solicitor to plead my case at this tribunal was called John F. Harrison but it turned out that he had prior engagements on the day of my hearing and so he in turn appointed another solicitor to represent me, Mr. Sanjit Thaliwal. I looked forward apprehensively to the day of battle against the medical profession in Birmingham once more but whatever the outcome of my appeal, I was submitting my manuscript to my three preferred would-be publishers the very next day when I did know the outcome. I prepared my case papers that were to be used in France in readiness to be used at the tribunal here as well. When Friday finally arrived, I was up and at

THE DARK SIDE OF THE BBC. A DISTURBING TRUE STORY.

'em early as usual and my solicitor arrived about an hour early to give me the final briefing. A room was set aside for the tribunal hearing and the members of the board arrived early as well so that they too could make final preparations. We entered the room set aside for the hearing at 2-30 pm and the hearing began. There were three members of the tribunal, the president Mr. R. B. Almond, the doctor Mr. C. Berry and the lay person Mrs. S. A. Riordan. I was asked to speak first (oh shit!) But I came across as plausible and confident. The consultant psychiatrist then made his case against me, followed by the social worker and then my solicitor picked them up on the key points and made his case in my defense. He did very well I thought and this was proved after two hours of deliberation and discussion when the tribunal came back with the decision that my appeal against section under the mental health was successful. I was a free man. Free to continue my mission to expose and eradicate the existence and miss-use of 'telepathy' again.

SOLDIERING ON.

THE DARK SIDE OF THE BBC. A DISTURBING TRUE STORY.

Time passed by pretty quickly from then on and I tried my best to keep out of trouble. All the trouble I did get into was obviously a direct result of the illegal activities of the BBC. But they would go around telling people that I was a criminal and probably had a copy of my criminal record to hand in order to further scandalize me. So I tried to keep myself to myself as much as I could but when every-one is linked to your thoughts and vision it can be impossible sometimes. In any case I had almost finished my book so I thought it time to write off to a publishing company as the next step to getting published. I went to the central library in Birmingham and got a list of all the publishers that might publish a book like mine. I ended up doing a mail shot to about fifty or sixty publishers then waited for the replies to come rolling in. I waited a week, then two, then after the third week the replies started coming in and I was hopeful. Unfortunately none of them were prepared to take on the BBC and I was let down time and time again. One publisher even recommended another publisher to me saying that they were a very combative publisher. This made me smile so I wrote to them but they eventually declined as well. Then there was a ray of light after the fourth week. A few more letters came with all refusals but one came from Janus Publishers Ltd. They said that they would publish my book but that I would have to pay £3,800 towards the cost. It was a new scheme they operated but it was my only hope. They invited me to send a copy of my book to them on paper and on disc. I did that and waited for a reply while I began saving money like mad. Eventually they wrote back to me saying

THE DARK SIDE OF THE BBC. A DISTURBING TRUE STORY.

that they would be honored to publish my book. I was overjoyed with that and rushed to get my book finished and up to date as best I could.

In the meantime, I had some terrible news from France. On 17th October 2003, I had a letter come from the court of human rights in Strasbourg saying that a committee of three judges had ruled my application as inadmissible because it did not comply with the requirements set out in articles 34 and 35 of the convention. What this meant was that they had ruled my application as inadmissible because I had not followed correct procedure and had not exhausted all possible remedies in the courts in the UK. Yet the convention clearly states that you must try to resolve your complaint as best you can in the UK, UNLESS you have a good reason not to be able to and I stated in my application that I could not resolve my complaint through the legal channels in the UK because the BBC were controlling everyone with their telepathy, the courts included, and making them operate against me. If the letter I got had said that my application had been made inadmissible because the BBC had landed at the courts and covered up my complaint, well at least I could have believed that. You can draw your own conclusions. In any case the court of human rights in Strasbourg, France are just as dirty as the rest of them.

When I phoned the courts after that to see if I could get my case reinstated, that same Steven Philips came on the line and he became quite nasty towards me when I was being polite and shouted at me telling me not to contact the courts again. Now with that sort of behavior from the head of division, you try telling me that the BBC were not involved there one way or another. In any

THE DARK SIDE OF THE BBC. A DISTURBING TRUE STORY.

case, it would be a tactical move they would have to make in order to continue with their cover-up.

I got some news of a completely different kind as well that year. Jane Henry who I had sex with once had become pregnant by me and was about to give birth to a baby boy. She was no good though obviously and abandoned the child after his birth in Heartlands Hospital. I went through the courts to try and get custody of him because I loved him as any father would but because of my phony mental illness I was refused and the child was put up for adoption. I went to the registry office in Birmingham with the senior social worker on the case and registered him as Anthony Kieran Henry and he eventually went to a nice family who I still keep in contact with now through the letter box scheme. I write to his adoptive parents twice a year and they write back to me twice a year. They changed his name to Dominic which I think is a nice name and he continued to grow into a good healthy boy.

I soon began to realize that my next door neighbor Lincoln really did have broadcasting equipment in his flat and when I was entering my flat one day with Jane, she looked over at Lincoln's door and said to me "yes that's where it's all happening, in there". I sort of knew what she meant and the next few days saw a rise in the 'voices' in my head. They were trying to get me to fight Lincoln saying that he was the "weakest link" as he is only small and skinny. It was driving me mad for days not knowing for sure so the 'voices' then switched from fighting Lincoln to breaking into his flat. I came close one night at about midnight but pulled away from doing it at the last second. After that close call, the voices must have thought "bloody hell, he really will do it" so

THE DARK SIDE OF THE BBC. A DISTURBING TRUE STORY.

sometime while I was out the next day, they came and took the 'machine' out of his flat and hid it somewhere else.

When I went to bed the next night, they really went to town on me to get me to really do the deed knowing that there was no way they could get caught out with the 'machine' hidden somewhere else.

In the run up to that night, I heard a message from a woman on the television saying "because he tried".

She was saying that because I tried to tackle the BBC in the human rights court and they must have had to physically go there to cover it all up, they were now going to really go to town on me in another attempt to cause me a break-down. So they began to broadcast to everyone about my sexual activities hoping to embarrass me to death. But I did not care, I had nothing to be ashamed or embarrassed about and if I did, it was their invasion of my mind that caused me to do it. So I soldiered on until I could not take anymore of the 'voices' in my head telling me to go and kick in Lincoln's door because he was dealing with all that sort of smut from his flat. I took a hammer from my tool box and went outside to his door at about two or three o'clock in the morning. With two powerful boots to his door, it burst open and in I went to smash up the machine. Of course it wasn't there so after a couple of minutes of searching with Lincoln shouting for me to get out of his flat, I had to concede and apologized to Lincoln as I left telling him to just phone the council and they will come and fix his door for free.

Of course he didn't, he phoned the police and they came in force within a matter of minutes. I wouldn't answer the door at first which was a bit stupid really because they kicked the door in within minutes and carted me off to

THE DARK SIDE OF THE BBC. A DISTURBING TRUE STORY.

stechford police station in hand-cuffs. I was charged with aggravated burglary which is a very, very serious charge so I made a no comment statement and was taken to court the next day. Everybody was against me again this time because I had nearly delivered a knock-out blow to the BBC and they were out for revenge. The first magistrate I appeared before was a snotty old woman who struggled to suppress her desire to send me to prison without bail and so I was carted off to winson green prison and I was as sick as a parrot. I was going to spend a long time in prison I knew awaiting trial and that is the last place a telepath like me would want to be. To make things worse, the BBC managed to bully their way into the prison hierarchy and get one of their 'machines' brought into the prison so that the inmates could bully me whilst I was there. I was held on remand for about a year and was subjected to mental abuse almost every day and night for the duration. Everybody outside the prison could hear in their heads what was going on every day as my thoughts and vision was left broadcasting to all around Birmingham day and night. The inmates were encouraged to bully me for my sexual exploits and masturbation was the key issue but it was all bollox because for all the time I was in prison, I did not masturbate once, yet everyone else was bang at it on a regular basis because there was no women to be had in jail. Also, all the verbal abuse that was directed at me was done through the windows and door when I was locked up and when I was outside my cell, all the big mouths stayed quiet. Typical of the cowardly conspiracy we were all involved in here in Birmingham.

With all that going on, I did not have any time to myself to prepare my defense and when it came to near trial time, my legal team did all they could to stop me from testifying against the BBC. In fact, on the morning of my trial, because I

THE DARK SIDE OF THE BBC. A DISTURBING TRUE STORY.

was not guilty of aggravated burglary because I had no intention of stealing or of violence, my barrister went into court and told the prosecution to add an alternative lesser charge of ordinary burglary which I had to plead guilty to. Another legal cover-up. So eventually I was going to be given a hospital order again and was transferred to meadowcroft hospital in winson green where I would await sentence. The BBC wanted me to fight all the way throughout my incarceration but I couldn't be bothered, I wasn't going to get into more trouble just to please them especially when it was them who got me in prison in the first place. Some people were stupid enough to think that I was a soft touch because I wouldn't fight until one of the patients in meadowcroft purposely bumped into me in one of the corridors. I let it go at first thinking he was a bit of a prick but when he did it again the next night I knocked the crap out of him in a matter of seconds. He scraped himself up off the floor and ran off to his room so I let him go. He would not try that again.

In the meantime, whilst I was in meadowcroft, I signed a contract with my publishers – Janus Publishers Ltd -.and paid them £2000 towards the cost of publication. The only trouble was that the BBC had landed there to stop the publication of this book. Again, it was a tactical move they would have to make in order to continue with the cover-up of their illegal activities. They changed the title of my book to The Dark Side of Television, 'Fact or Fiction', then they changed all references to the BBC to the CBC and changed the names of people and places in my book. Then when they really had Janus Publishers Ltd. Eating out of their hand, they changed the title again from 'fact or fiction' to 'belief is a sign of grandeur'. They were really taking the piss then by saying that I had delusions of grandeur for thinking that the BBC were ever interested in me.

THE DARK SIDE OF THE BBC. A DISTURBING TRUE STORY.

Cheeky fuckers! It was S. K. Leung from Janus publishers that did the deed against me and he ripped me off for two grand. I heard from a reliable source that he later gave the business to his daughter Jeanie. Anyway, the two offending front covers intended for my book are also kept on file.

I heard that my flat had been burgled whilst I was locked up so after a couple of weeks in meadow croft, I was walking on the hospital grounds with two male escorts, when I decided to run off and check my flat out for myself. The two nurses escorting me chased me for a little while calling my name as they ran after me but I was too fast for them and was long gone within minutes. When I got to my flat, I discovered that I had indeed been burgled and everything was gone except for the odd pieces of furniture. I was wounded but soldiered on despite my circumstances. I linked up with my mate Pete Flynn and had a few drinks and a few laughs before I would hand myself in to the police. My Dad told me that I was on the news on heart fem. radio. They said that I had escaped from a secure mental hospital but I was not thought to be a danger to the public. Yet when I gave myself in and went to court for sentence, they said that I was a danger to the public and should be sent to a high secure hospital where I can only get out when the home office or a mental health review tribunal say I can. I was really in the shit now I thought.

So I spent over a year in reside hospital trying to keep my head together. But it wasn't all that bad in there really. I got on well with Dr. Clarke my consultant and with the rest of my team, I liked them and I think they liked me and I was eventually freed by a tribunal and given a lovely little flat in Birmingham. They said I could not go back to my old flat because of the risk I posed to Lincoln. In any case I settled in to my new surroundings before the BBC were on me again.

THE DARK SIDE OF THE BBC. A DISTURBING TRUE STORY.

This time they gave a machine to my next door neighbor called Mark and he was next to try and break me down with all that telepathy again. I heard people mention his name as the new one to control the people of Birmingham with telepathy. But he was another big pussy like the last two and he didn't show himself for over a year from when this new phase started. Then it was only a quick glimpse when he did.

Unfortunately for me, I managed to break my back in an accident at Pete's house just before Christmas 2006. I ended up in hospital in a wheel-chair at first, then on crutches before I could walk again on my own two feet. This did not stop all the telepathy though and people continued to try and make me fight in the streets as usual which was pretty shitty. Then a new guy moved into the flat underneath me and after a couple of months he was involved into the plot against me. He would be banging walls and talking to my thoughts through the ceiling and generally trying to harass the shit out of me. I bumped into him a couple of times in the street and warned him that I would knock him out if he continued with his actions but that did not stop him probably because that Mark kept putting him up to it. What ever the case, I ended up going into his flat at four in the morning with a young lad I was friendly with and knocking the shit out of him. I was drawn into crime yet again and this new guy told the police on me and I was arrested later on that same morning. I was charged with burglary and assault causing actual bodily harm and remanded back into prison with no chance of bail. The rest is history repeated. I went through the big legal cover-up once again with everyone from my own solicitors to the crown court judges trying their best to implement the medical cover-up at my trial and so in the end I had to plead guilty or face the consequences. I pleaded

THE DARK SIDE OF THE BBC. A DISTURBING TRUE STORY.

guilty and ended up back in reside for only a short three month stay before I was given another lovely little flat about a mile up the road from my last one when I was released. My mate Pete Flynn helped out with the decorating and the furniture transportation and so I settled in for the long stay this time as I was determined not to get myself arrested again due to this bbc conspiracy against my wishes. I busied myself finishing off the writing of this book before publishing it myself without relying on anyone else again who could scupper my plans to publish. Now the whole world has something readable to reflect upon for the next thousand years with the reading of this book. Unfortunately, the BBC had upped their game for a while now and were concentrating on assassinating my character. They wanted me to fuck me up bad now after I had taken them to the court of human rights to teach me a lesson so I would never do anything like that again. So they allowed the scum of the BBC to come into action with their sleaze stories and their dirty minds. Of course when you have perverts that stalk and spy on others, it always comes down to sex. So they busied themselves stopping me from having any love or sex with another woman so that I would be left cold and alone while they told the people sleazy stories about me and my sex life initially, then eventually, it was always about masturbation and always it was made out to be something dirty. They used people to call me nasty names all the time and try to degrade me with words like wanker and pussy coward and pervert as though these words were true and always they lay in wait for me to masturbate so they could broadcast it to everyone to degrade me even more. Well they were gonna have a long wait, and always these BBC scumbags that weren't good enough to make it onto television and pervert mark (as I now called him) as their cowardly leader

THE DARK SIDE OF THE BBC. A DISTURBING TRUE STORY.

would translate everything from their dirty point of view and lie cheat and exaggerate to make me look bad. But everybody masturbates, whenever, wherever, it is human nature, so how can the BBC or anybody else think masturbation is dirty, when they all do it themselves?

In any case it did get me mad sometimes and when I get mad, all these hypocritical tossers would disappear like rats fleeing a sinking ship. I did not want to spend the rest of my life surrounded with all this shit so I made some enquiries to the legal profession and discovered that I could represent myself in court and sue the BBC myself without having to rely on another solicitor that would be turned against me. So on 21/02/2011, I filed my first lawsuit against the BBC for £5,000,000 for invasion of privacy. I filed it in Birmingham County Court along with a copy of my self published book, five witness statements and a copy of my petition against the existence and misuse of telepathy and went home with a tinge of hope in my heart that this perverted activity would soon be at an end. I waited and waited. Eventually I went back to the Courts to find out what the delay was and after some investigations on the courts computer, I was told that there was no record of my lawsuit ever being filed. Wow. I was gob-smacked. This meant that the BBC had infiltrated the Courts and stolen my paperwork as a means to stop me from suing them and that someone who worked for the Courts had allowed them to do it. I asked myself then, how can anyone have any faith in the British judicial system now after that little bit of news. But still no-one complained. They were all brainwashed by now like a herd of cattle being herded along by the low lifes of the BBC.

I had no other option than to file a second lawsuit against the BBC and did just that on 3/3/2011. I did not enclose a copy of my book and the other documents

THE DARK SIDE OF THE BBC. A DISTURBING TRUE STORY.

this time as it would have meant that I would lose £10 every time my book went missing but I was clever enough this time to make sure I got the Courts date stamp on my copy of the lawsuit as proof of receipt. Now this case could not go missing and had to go before a judge but my story has to end here at this point as I am publishing this book now with an American publishing company called Authorhouse. You can find out what happened next in my second book due out in or around October 2012. Finally, as a last thought and just to let you know that the BBC rake in one point four billion pounds a year approximately off you the British TV license payer and I wonder how much of your money they spend on producing the machines and the technology that they use to invade the minds of ordinary people like me and you. The mind boggles.

If you have any views personal experience or comments you would like to raise on the contents of this book, especially if you live in Birmingham or London, please feel free to email me at anthonyhickey66@yahoo.com

END OF PART ONE

THE DARK SIDE OF THE BBC. A DISTURBING TRUE STORY.

(LOOK OUT FOR 'THE DARK SIDE OF BBC TELEVISION PART 2'.
END GAME)

Printed in Great Britain
by Amazon